T0130900

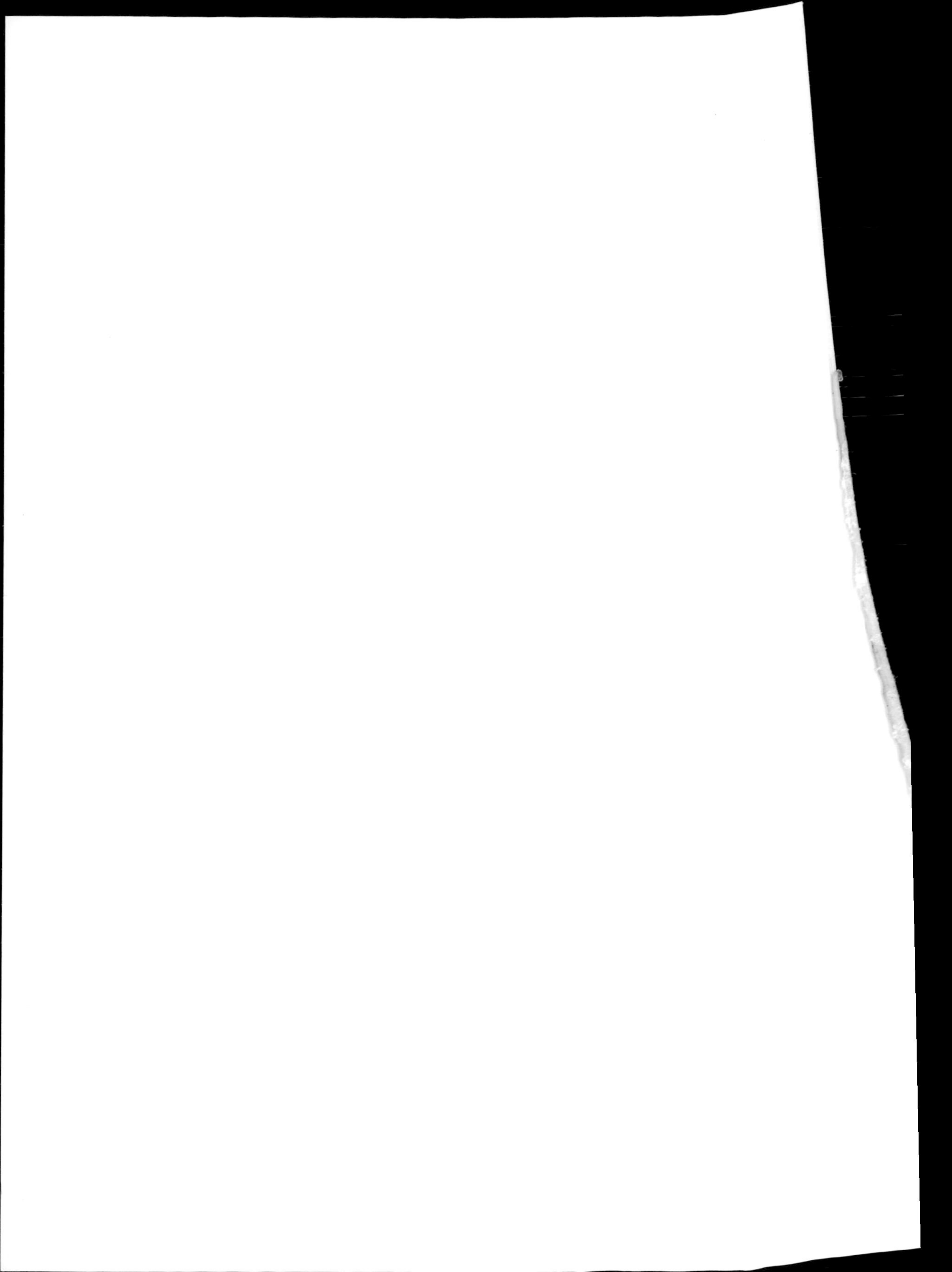

Cut-and-Paste Genetics

Cut-and-Paste Genetics

A CRISPR Revolution

Sahotra Sarkar

ROWMAN & LITTLEFIELD

Lanham • Boulder • New York • London

Published by Rowman & Littlefield
An imprint of The Rowman & Littlefield Publishing Group, Inc.
4501 Forbes Boulevard, Suite 200, Lanham, Maryland 20706
www.rowman.com

86-90 Paul Street, London EC2A 4NE

British Library Cataloguing in Publication Information Available

Library of Congress Cataloging-in-Publication Data

Names: Sarkar, Sahotra, author.
Title: Cut-and-paste genetics : a CRISPR revolution / Sahotra Sarkar.
Description: Lanham : Rowman & Littlefield Publishers, [2021] |
Includes bibliographical references and index.
Identifiers: LCCN 2021015619 (print) | LCCN 2021015620 (ebook) |
ISBN 9781786614377 (cloth) | ISBN 9781786614391 (epub) Subjects: MESH:
Gene Editing—ethics | CRISPR-Cas Systems | Clustered Regularly
Interspaced Short Palindromic Repeats | CRISPR-Associated Protein 9 |
Gene Editing—history | Genetic Diseases, Inborn—therapy
Classification: LCC QH440 (print) | LCC QH440 (ebook) | NLM QU 550.5.G47 |
DDC 576.5—dc23
LC record available at https://lccn.loc.gov/2021015619
LC ebook record available at https://lccn.loc.gov/2021015620

To my daughters, Kajri and Malini, part of the last generation born without the specter of germline intervention before birth

Contents

Preface

Wired magazine is not known for understatement or subtlety. So the rhetorical excess of the cover of *Wired* 's August 2015 issue should not be regarded as surprising. "PLAY GOD," the cover proclaimed. "No hunger. No pollution. No disease. And the end of life as we know it. The Genesis Engine. Editing DNA is now as easy as cut and paste. Welcome to the post-natural world." What was all this excitement supposed to be about? "We have the power to quickly and easily alter DNA," *Wired* 's article explained. "It could eliminate disease. It could solve world hunger. It could provide unlimited energy. It could really get out of hand." Some readers were probably surprised that all this hyperbole was generated by two small molecules, a single RNA segment part of which matched a targeted DNA sequence, along with a protein that cuts DNA. The technique, dubbed "CRISPR" after the name of a microbial genomic structure in which it had first been found, had only been invented in 2012. By 2015, it was being touted as the greatest advance ever in biology. *Wired* was not only wired; it was pumped up.

Welcome to the world of CRISPR, dominated by good science but also soaring rhetoric and unprecedented venture capital. In 2005, in part, the *Wired* writers and editors were justified. In 2020, the two biologists who had made the most significant contributions to the development of CRISPR, Jennifer Doudna and Emmanuelle Charpentier, were jointly awarded the Nobel Prize in Chemistry. The honor was expected.

Wired's hyperbole understandably generated widespread derision on social media. "CRISPR makes the Most Interesting Man in the World more interesting," was one #crisprfact on Twitter. "If you apply CRISPR to Joyce's Ulysses, you discover it is really a short story by Hemingway" was another. More wittily: "'You must be the change you want to see in the world'.—CRISPR."

Hype aside, here are some simple facts: CRISPR will not eliminate disease. It will not solve world hunger which is caused more by the disparity of power between the overfed and the hungry than it is by biological limitations of food production. It is hard to imagine how CRISPR could produce unlimited energy (besides contradicting the conservation laws of physics). And so on. To the best of my knowledge, CRISPR researchers have not made any of these claims. That credit must go to the editors of *Wired.*

Nevertheless, CRISPR is the most powerful technology that biology has ever produced. The *Wired* article correctly points out that it has made gene editing easy. It has made it cheap. CRISPR can be used to target any gene in any organism. It can alter that gene to any other version we want. The technology is getting better. It really could eliminate many genetic diseases for which there is no treatment. It is also true that it could get out of hand. But how, and to what extent? These are the issues that this book is about. What motivates it is the expectation—and worry—that CRISPR will be used to consciously alter the genetic and thus the evolutionary future of the human species. *This* is an eugenic project, as inevitable as it demands caution. CRISPR has the potential to induce many revolutions, from a new class of genetically modified organisms (GMOs) for use as food to the elimination of insects that transmit disease. While all of these possibilities will find space in this book, its major focus will be on a potential eugenic revolution: altering the human gene pool for all future generations. Like Joe, the fat boy in the *Pickwick Papers*, "I wants to make your flesh creep."

The eugenic project has a long and chequered history going back to Francis Galton, Darwin's cousin and a towering figure in modern statistics, who coined the word "eugenics" (meaning well-born) in 1883. In the early part of the twentieth century, its proponents included stalwarts of the Left and the Right, liberals and conservatives, environmental conservationists, and suffragists. Its reach was global: the Europeans took it to their colonies, the Zionists brought it to Palestine, and Japanese opted for homegrown concoctions. In the North, the project was gradually but inexorably appropriated by racist ideologues typically operating within the political mainstream. In the United States, it led to forced sterilization of women, a practice that continued legally until 1983 when Oregon finally repealed the last state eugenics law. (But involuntary sterilization of prisoners without proper consent continued even after that, with 148 cases in California between 2006 and 2010.) The US sterilization laws made a vivid impression in the minds of a young Austrian in the 1920s. His name was Adolf Hitler and he went on to found the most vicious regime of the twentieth century.

The Nazi abuse of biology is well known. It included the eugenic elimination of those deemed undesirable including Jews, Romanos, and homosexuals. The end of the Nazi regime and exposure of its eugenic and other medical atrocities led to widespread revulsion and a sea change in attitudes toward ethically acceptable biomedical practices in the North. Subsequently, these transformed attitudes led to the adoption of stringent protocols for physicians' conduct that emphasized patient autonomy and physician responsibility. These protocols rapidly diffused through the world. Eugenics took on a dirty—even scary—association that it has never been able to shake off even though many of the arguments of its proponents were never scientifically refuted. To call a practice eugenic was to brandish it with the legacy of the Nazis. In the 1950s and 1960s, it became hard to find self-proclaimed eugenicists. But sympathy for eugenics never disappeared; it simply went underground to wait for more sympathetic circumstances.

Meanwhile, the same post–World War II period saw the emergence of modern molecular biology, especially molecular genetics, which rapidly spawned biotechnology tools of unprecedented power. It became plausible to suggest modifying human genes that were implicated in disease. However, the technology that could be deployed for this purpose, known as recombinant DNA and invented early in the 1970s, was not limited to modifying only such genes. It enabled editing the genes of all species and the creation of GMOs that contained the genes of other species. Meanwhile, the polymerase chain reaction (PCR) technology developed by Kary Mullis in 1983 allowed rapid copying of DNA to generate vast quantities of identical sequences from a single original template. That also had technological consequences that we are still reaping, from the mass production of life-saving drugs to the identification of disease-causing microbes. (If you have been tested for Covid-19 in the North, very likely it was a PCR-based test.)

These developments occurred within a social context of pervasive genetic reductionism, a widespread belief within biology and beyond that genes alone are the most important causal factors responsible for traits, both physical traits such as skin color and height, and mental and behavioral traits such as intelligence and components of temperament. Biology in the twentieth century came to be dominated by genetics. Genetic reductionism along with the dominance of genetics over the rest of biology contributed to the launch of the controversial Human Genome Project (HGP) in 1990, a crash program to sequence the three billion nucleotide bases that comprise the human genome. Sequencing would be "blind" with no attention paid to whether a particular DNA segment harbored functional parts. (These functional parts were then believed to comprise less than 10% of the genome.) The expectation was that the sequence would provide powerful predictions about a person's traits and transform medical practice. We were supposed to get a personalized medicine based on each of our unique genomic DNA sequences. Even though the proposed project had many scientific and other critics, its proponents prevailed and the project proceeded in spite of its enormous cost. In the United States, the federal government spent at least three billion dollars.

The eugenic implications of the HGP were acknowledged from the beginning as were other potential social ramifications of the project. A significant fraction of the project's budget was assigned to studies of the ethical, legal, and social implications (ELSI) of the human genome. These studies were supposed to provide guidance on how society should address problems that were expected to arise once the sequence became known. Though the HGP was completed ahead of schedule in 2000, by and large, the anticipated social problems never materialized. The most important concern had been genetic discrimination, especially the potential use of genetic data by insurers to deny health insurance on the ground that someone's DNA sequence indicated susceptibility to some disease or disability. In the United States, that concern was addressed through legislation, in particular, the Genetic Information Nondiscrimination Act of 2008 which established strong safeguards for genetic privacy. (Of course, this concern did not arise to the same extent in more civilized societies where access to medical care was acknowledged as a basic human right.)

Most anticipated problems from the HGP did not materialize because, scientifically, the sequence failed to live up to its hype. Whether it be diseases or other traits, the DNA sequence has proved to be a poor guide to the biology of individuals except in the rare cases where single (or very few) genes strongly affected a trait. But these were a tiny fraction of traits and almost all of them were well known long before the HGP. Geneticists had been studying these traits since the beginning of the twentieth century. If we restrict ourselves to diseases, these are traits such as color blindness, hemophilia, sickle cell disease, myotonic dystrophy, cystic fibrosis, and Huntington's disease, all known long before anyone dreamed of the HGP.

Skeptics of genetic reductionism—and of the HGP itself—are justified in feeling vindicated in their criticism of the HGP. As the critics had emphasized, traits are the result of the complex process of embryonic development from a fertilized egg to an adult organism. Genes are a critical resource for this process, and for some traits the most influential ones, but they are only one of many materials that are entangled in the biology of an organism. Most importantly, the developmental process is history dependent, relying on the presence of the correct physical pieces and interactions at each stage. This perspective of *contextual developmental construction* of organisms pervades this book and stands in sharp contrast to the facile genetic reductionism that animated the HGP.

Because of its implicit reliance on genetic reductionism and its failure to embrace developmental complexities, the HGP also failed to deliver on any of its medical promises. A *Scientific American* report from 2010 concluded that it had made no tangible difference to medicine. At the time prospects for eugenics seemed dim and worries about it far-fetched. But all that changed in 2012 with the development of the CRISPR technology for gene editing. This technology was indirectly a result of the HGP as we shall see later in this book: its creation depended on the fast sequencing techniques, bioinformatics, and other technologies spawned by that Project. By making precise gene editing easy, CRISPR has revived eugenics and other dreams of the genetic reductionists even though the biological complexities of the developmental process remained the same.

For medicine and eugenics, the advent of CRISPR does make a difference. It makes a strong case for genetic intervention for those diseases that are strongly affected by a single gene. These include several diseases such as cystic fibrosis, hemophilia B, Huntington's disease, myotonic dystrophy, and, sickle cell disease, all identified in 2017 as medical priorities for gene editing by the US National Academies of Sciences, Engineering, and Medicine. When gene editing is confined to *somatic* cells (those that would not be the source of future generations), the Academies supported gene editing so long as the techniques used were shown to be safe and effective. With the restriction to somatic cells, the edited genes would not be passed on to a person's children. Such use of CRISPR techniques would be ethically no different from any other kind of intrusive medical intervention such as many forms of surgery so long as standards of efficacy and safety are maintained. This is the most uncontroversial medical promise of CRISPR. However, compared to germline editing, it is difficult: the molecules that form part of the CRISPR toolkit would have to be precisely delivered to the intended cells and only to those cells.

The real power of CRISPR comes from the fact that we could just as easily intervene in the germline as in somatic cells. This is what generates the potential for eugenics. We can potentially eliminate a host of genetic diseases from the human population including the ones that were mentioned earlier. This is technologically simpler than somatic cell gene editing because the molecular toolkit can be injected into an embryo at the single cell stage (what is called the zygote). It should come as no surprise that germline gene editing has already been attempted, though so far only in rogue experiments carried out in China in 2018. Though the biologist who carried out this experiment, He Jiankui, was roundly condemned by colleagues globally and subsequently imprisoned by Chinese authorities, it is far from clear that what he attempted was ethically problematic for its eugenic aspect or only because he seems to have been in violation of safety and other regulations about the treatment of human subjects. Many others will follow Je, possibly in secret; in Russia, Denis Rebrikov has already informed the world (and the authorities who would have to give permission) that he wanted to carry out disease-implicated germline editing. So far, he has not been able to get the necessary permissions to proceed with his experiment. He and Rebrikov are only the first two of likely many researchers who will embrace eugenic measures to eliminate the disease.

Thus, CRISPR-based easy gene editing has made tangible the possibility of eugenics through germline manipulation in a way we have never encountered before. Gone are the days when eugenics required sterilization or slaughter, or even restraint from reproduction. Rather, eugenics would consist of allowing parents to produce embryos that would be assayed for the presence of suboptimal genes. These genes would then be modified (edited or even replaced) using CRISPR technology. Though not yet perfect, the process is simple, accurate, cheap, and flexible (insofar as any gene can be fixed). The technology is getting better every day. The germline-edited embryos would then be implanted in the mother. About thirty-six weeks later, a genetically transformed baby would be born. The whole process would only be a slight wrinkle on how babies are already produced using *in vitro* fertilization. As CRISPR and associated technologies improve, the process will become increasingly less intrusive and better. Fewer embryos will be needed for implantation. More and more genes will be edited.

Whether we like it or not, this new eugenics is inevitable. It already has a vocal constituency: families and friends of those who suffer from debilitating genetic diseases caused by single mutant genes and with no adequate treatment. CRISPR-based gene editing is already being used to treat these genetic diseases in trials that are underway in several countries. So far, these trials have been restricted to gene editing in somatic tissues. As far as is publicly known, germline intervention has only been attempted in the rogue Chinese experiments mentioned earlier. But, as we will see later in this book, there may be no good reason to avoid germline intervention in situations when a single mutant copy of one gene causes a disease. Thus, CRISPR-based germline intervention into Huntington's disease and myotonic dystrophy, and other such diseases will likely begin very soon and usher in an era of conscious human intervention into the germline, changing the gene pool of our species for the future.

But, after these simple cases, we find ourselves on a slippery slope. What about diseases that require two copies of the faulty gene to be manifested? These include cystic fibrosis, hemophilia B, and sickle cell disease. Should we intervene in the germline of an embryo with only one copy of the faulty gene knowing that that embryo would not suffer from the disease but would still be able to pass the gene on to future generations?

What about diseases or perhaps, conditions, such as hypertension that may be influenced by genes but can be relatively easily managed through ordinary medication? What about hereditary deafness which some parents prefer to transmit to their children? What about mild attention deficit hyperactivity disorder (ADHD), given that many of those who have that condition accept it as natural variation? Should parents opt for germline intervention for all of these cases? Should society permit it? The advent of CRISPR technology has emboldened a cadre of "liberal eugenicists" who consider themselves liberal because they are adamant that germline intervention decisions are left to individual parents rather than allow society to be involved in any way in these decisions. Coercion in any guise is anathema to liberal eugenicists. Most of them would urge the freedom of parents to opt for germline intervention for all the cases just mentioned: hypertension, deafness, ADHD, and many other such cases in which there is a partial genetic etiology. CRISPR has also emboldened "moderate eugenicists" who accept some social intervention in reproductive decision with the goal of preventing genetic diseases. They see this form of eugenics as a public health measure similar to mandatory vaccination, quarantine, or lockdown in the face of an infectious disease pandemic.

So far, we have only considered diseases or conditions which are generally socially judged as undesirable though, perhaps, without full justification in some cases such as hereditary deafness and mild ADHD. But CRISPR-facilitated eugenics need not be confined to these situations. Most liberal eugenicists do not limit their dreams to the elimination of diseases. They also promote genetic enhancement of desirable traits, focusing on traits that range from physical prowess to cognitive abilities. At least in the North, there seems to be a widespread belief that the move from germline editing for eliminating disease to germline editing for genetic enhancement crosses some deep moral divide. Whether this is so, remains far from clear. What is clear, though, is that genetic enhancement, even if confined to somatic tissues, would raise troubling questions of equity. Enhancement is not a medical choice because it is not just about restoring and maintaining normal healthy functioning. It is about transcending normal functioning. Because enhancement must go beyond the expected normal range it is unlikely that its costs would be covered by the state or private insurance. Would it, therefore, become yet another advantage enjoyed only by the wealthy and thus serve to increase already pervasive forms of social inequalities?

While ethicists have written volumes debating the morality of genetic enhancement, the disagreements are nowhere near resolution. Indeed, it is not even clear that much new insight has emerged from these ethical discussions during the past decade. What has been missing from discussions of contemporary eugenics is a hard look at what the new scientific developments say about its credible prospects. We have not had our version of J. B. S. Haldane's *Heredity and Politics* from 1938, an incisive intervention by one of the world's most prominent geneticists into the debates over eugenics during the Nazi era. Haldane provided a decisive technical critique of eugenics that was beyond scientific refutation. For that matter, we have not even had a balanced exploration of scientific possibilities such as Neil Holtzman's *Proceed with Caution* from 1989 which appeared just as the HGP was being initiated. As we shall see later in this book, much of the ethical debates about genetic enhancement proceeds in a world of make-believe science. There is no prospect of genetic enhancement in the foreseeable future and there are deep biological reasons why it may never be feasible. CRISPR will not produce designer babies. Meanwhile, in *Biotech Juggernaut* (from 2019) Tina Stevens and Stuart Newman have provided a stinging critique of the social role and power of biotechnology, including CRISPR, on our daily lives.

Indulging in flights of fancy about genetic enhancement would not be such a bad thing were it not for the fact that they are a waste of time and resources at a time when we need to make immediate social policy decisions on how to regulate inevitable eugenics in a CRISPR-enabled world. As the rogue experiments in China have forcefully brought to our attention, we need policy guidelines of when, if at all, it is permissible to edit the human germline and who should be empowered to make that decision. This need is widely acknowledged and various national academies and other prestigious bodies, as well as the World Health Organization, have set up one committee after another to address these issues. These committees have routinely provided bland discussions of ethics and policy, not much different from the ethicists' contributions, but typically have avoided making concrete recommendations about what should be done now.

The first report to break this mold to some extent was a 2020 contribution from the (United States) National Academy of Sciences (with participation from the US National Academy of Medicine and the Royal Society of London). That report was produced by an International Commission on the Clinical Use of Human Germline Genome Editing assembled in the wake of the rogue Chinese experiments. It tamely—but correctly—concluded that existing technologies, including CRISPR, had not yet been demonstrated to be safe enough for clinical use. Beyond that, it limited its discussion of the use of human germline editing to disease prevention. It recommended the initial restriction of the use of this option, after safety has been demonstrated, to cases in which a single copy of a gene causes severe disease. But it was short on the biological reasons why this is a wise choice.

Nevertheless, that is the discussion we should be having and much of the ongoing discussion of the ethics of enhancement is an unfortunate distraction. Science fiction may be fun in literature, philosophy, or sociology classes but is not quite so relevant in hospital wards. We should worry about the possible elimination of genetic diseases, and possible unintended consequences of those efforts, and not about the supposed potential for a genetic enhancement of intelligence.

This book will set the CRISPR-induced prospect of eugenics in its historical, philosophical, and scientific contexts. By now, the history of eugenics and the horrors committed in its name are widely recognized. So, the first part of the book will focus only on those parts of that history that have relevance to us today. We will then see how and why a failure of molecular medicine has renewed the prospects for eugenic germline intervention as our best available response to genetic diseases. These prospects have become much better because of the two developments we have been discussing: liberal eugenics which promises that this brave new eugenics will not permit repetition of the horrors of the past; and CRISPR which provides the technological prowess.

Let us turn briefly to what CRISPR is before we head into the book. "CRISPR" is short for clustered regularly interspaced short palindromic repeats of DNA sequences that form an array including variable bits of DNA called *spacers* between the repeats. These arrays are found in many bacteria and archae which are all single-celled organisms called prokaryotes and distinguished by the fact that they do not have a nucleus separated from the rest of the cell. The CRISPR arrays help defend the cells from invasions by viruses and other pathogens (as we will see in the fourth chapter). These invaders are recognized using not the CRISPR sequences themselves but by the spacer sequences that lie between the CRISPR repeats. Now, next to each CRISPR array are DNA sequences that specify CRISPR-associated (or Cas) proteins; some of these, especially one called Cas9, are very good at cutting DNA.

The CRISPR gene-editing technique uses an RNA sequence that plays the role of the spacer sequence in nature along with Cas9 (or a similar protein that cuts DNA). The RNA recognizes the targeted gene for editing and Cas9 cuts it. Then a corrected (or "edited") DNA string is inserted at this cut. There is an irony here: CRISPR technology does not use the CRISPR sequences themselves, that is, the sequences in the palindromic repeats. Rather, the name has stuck because of where the technology originated. (In this book, "CRISPR" will be used to refer to this technology in general.)

Because this technology can be used to modify the genomes of any species we choose, CRISPR is also being used to edit the genes of scores of commercially valuable species. The results are then being patented and acquired by a wide variety of biotech industrial corporations. Thus, CRISPR is also very big money. In 2018, market analysts estimated that the CRISPR market would be worth US$ 5.3 billion by 2025. CRISPR first became a business with the yogurt industry, even before its use for targeted gene editing. The yogurt industry was understandably interested in the resistance to pathogens of bacterial strains used to make yogurt and, thus in CRISPR arrays that had spacers that provided this resistance. Introducing a CRISPR array that generated resistance to any potent virus in a bacterial strain that produced a popular yogurt type could save millions of dollars in the yogurt production process as the industry happily found out.

The monetary opportunities of CRISPR have been recognized from the first days of the creation of the technology. After the advent of CRISPR-based gene editing using Cas9, in the United States, the University of California at Berkeley and the Broad Institute at MIT and Harvard waged a years-long patent battle in the courts. The latter eventually prevailed in the United States though not in the European Union. The economics and politics of CRISPR are important and interesting in their own right but will receive only glancing attention here. (For an entry into those issues Stevens and Newman's *Biotech Juggernaut* is recommended.) Let us turn to Eugenics, USA.

Chapter 1

Breeding a Perfect Society

> "The commonwealth is greater than any individual in it. Hence the rights of society over the life, the reproduction, the behavior and the traits of the individuals that compose it are, in all matters that concern the life and proper progress of society, limitless, and society may take life, may sterilize, may segregate so as to prevent marriage, may restrict liberty in a hundred ways."
>
> –Charles Benedict Davenport, 1911,
> *Heredity in Relation to Eugenics.*

BEGINNINGS: DAVENPORT AND EUGENICS IN THE UNITED STATES

In 2008, the Cold Spring Harbor Laboratory in Long Island, New York, one of the premier biological research institutes of the United States, published a volume, *Davenport's Dream: 21st Century Reflections on Heredity and Eugenics*. This laboratory had been set up in 1904 with funding from the Carnegie Institution in Washington. Its first director was Charles Benedict Davenport who worked hard to raise the money needed for its secure establishment and persistence. In 1903, Davenport was one of the first researchers in the United States to embrace the new Mendelian genetics, rules about the transmission of genes from one generation to the next that Mendel had established in the 1860s but had been ignored until 1900. Davenport made a few contributions to studies of the inheritance of pigmentation in humans, in particular, inheritance of eye color, a trait that he thought to be controlled by a single gene. We will encounter his work on eye color again in the seventh chapter of this book.)

Davenport is an apt beginning of our story because he was also an ardent eugenicist, having been converted to that cause by Galton, who had invented the word and created the field in 1883, and his disciple, Karl Pearson, both of whom Davenport had met during visits to London in the early 1900s. We have already encountered Galton in the Preface. Though he is not widely known by the general public, his influence on a wide variety of subjects from statistics through biology to psychology was immense. We will encounter him repeatedly as this book progresses. However, today, he is mostly associated with the promotion of eugenics and, in particular, a program of improving the human stock by encouraging increased breeding by those supposedly with more desirable qualities, especially intelligence.

Like Galton, Davenport was concerned with breeding a better stock of humans by encouraging the spread of better genes or, as Davenport preferred, better *genotypes*, that is, individuals with a superior complete set of genes. But, in contrast to Galton, Davenport's focus was more on the elimination of undesirable genes that, according to him, were increasingly being brought into the United States by new immigrants including Greeks, Hungarians, Irish, Italians, Jews, Poles, and Serbians. Davenport adhered to every class, gender, and racial prejudice of his day and continually obsessed about the sexual behavior of undesirable groups. For example, in his view, should sterilization become necessary for eugenic purposes, castration in males was preferable to vasectomy because it was also supposed to depress sexual desire and promiscuous behavior, and not just the ability to reproduce. When it came to sex, Davenport was very interested in what others did in the privacy of their bedrooms.

Nevertheless, Davenport was scientist enough to realize that, in the 1900s, there were as yet insufficient data to understand the genetic basis for most human traits. He became determined to collect those data and put them to proper eugenic use. In 1910, he convinced Mrs. E. H. Harriman, a wealthy widow with a railroad fortune, to fund a Eugenics Records Office devoted to gathering the data that he needed to frame properly scientific eugenic policies. It was set up next to the Cold Spring Harbor Laboratory in Long Island to ensure full collaboration between the two institutes. (The quotation with which this chapter starts comes from Davenport's 1911 book, *Heredity in Relation to Eugenics*, the publication of which was timed to promote the new institute. A facsimile of this book forms part of *Davenport's Dream.*) Over the years, the Eugenics Records Office, which lasted until 1939, trained and sent out hundreds of workers to collect data on the distribution of supposedly undesirable traits in targeted populations such as albinos in Massachusetts, the insane in a New Jersey hospital, and the feebleminded in a school, also in New Jersey. (Since the nineteenth century, "feeblemindedness" had been identified as a distinct trait indicating mental deficiency and was supposed to be correlated to a host of other undesirable traits including various forms of criminality.)

Davenport's data needs were very specific. In the 1900s and 1910s, genes were abstractions; their existence was inferred from family trees by the distribution of traits. The framework for making such inferences went back to the work of a Silesian monk, Gregor Mendel, who formulated the fundamental rules of genetics in the 1860s but was, as we noted earlier, ignored till 1900. Around that time, his work was finally appreciated by biologists and its breathtaking implications generated avid followers such as Davenport. Mendel had laid bare the principles of heredity that biologists since Darwin had been searching for, until then, with no success.

According to Mendel, each visible trait of an organism (these are called its *phenoytpic* traits), if it is due to a gene, would be influenced by two genes or *alleles* (which are different versions of a gene). If these alleles are alike, the organism is *homozygous* and the phenotypic trait will manifest itself in two types corresponding to each of the two possible homozygotes. If the two alleles are different, the organism is *heterozygous* but, according to Mendel, would look like one of the homozygotes. This phenotype is called *dominant*, the other apparently less powerful trait is a *recessive*. The corresponding genes (or alleles) are also similarly called dominant or recessive. (There are many exceptions to this rule of dominance but that is not relevant to us right here.)

An example from Mendel's experiments will help explain this convoluted terminology of genetics (and we will need it later in this book). Mendel observed that the color of pea seeds could be green or yellow: these are the two phenotypes. Seed color is specified by a single gene. Here, "gene" is replaced by "locus" in modern genetics indicating that what we are talking about is a position on a chromosome where a version of that gene (an allele) is specified. That genes, that is, loci are located on chromosomes and placed in a linear order like beads on a string only became established in the 1910s. (Thus "gene" sometimes means "locus" and sometimes "allele"; nowadays it also sometimes means the DNA sequence specifying an allele. These ambiguities sometimes make the language of genetics opaque but we must learn to live with this problem.)

Now, peas have paired chromosomes, like humans. Organisms with paired chromosomes are called *diploids*. Only about half of the flowering plant species are diploids. The other half have chromosomes occurring in quadruplets, octuplets, and so on—they are called *polyploids*. Humans are diploid and have forty-six chromosomes. Everyone has twenty-two paired chromosomes which are numbered based on their size. Females have a twenty-third pair of two *X* chromosomes. Males don't. Instead, they have one *X* chromosome inherited from their mothers and a tiny *Y* chromosome with very few genes from their fathers.

Returning to Mendel's experiments, there are two alleles that can influence seed color in peas. There is one allele for green color. Let us call it the *green* allele. (Throughout this book, as in much of recent biology, italics will indicate we are talking of an allele.) There is also a *yellow* allele. A plant with two *green* or two *yellow* alleles is a homozygote; as we would expect, these plants will have green or yellow seeds, respectively. (Notice that we don't italicize phenotypes.) However, some plants would have one *green* and one *yellow* allele. These are the heterozygotes. Now, green is the dominant phenotype. Dominance means that these heterozygotes will have green seed color as their phenotype. The *green* allele is dominant over the recessive *yellow* allele.

As we will see in the seventh chapter, Davenport, working with his wife, Gertrude C. Davenport, who was also a biologist, reported that a single pair of genes (that is, alleles) was responsible for eye color, and brown color was dominant over blue. (This was one of his few genuine contributions to genetics and, even here, as we shall see, the observation he reported is not universally true of humans.) The phenomenon of dominance shows that there is an important difference between the visible phenotype and the pair of genes or *genotype* responsible for it. Thus, a brown-eyed person could still be carrying an allele for blue eyes: there are two genotypes (*blue-blue* and *brown-blue*) corresponding to the single phenotype of brown eyes. The genes are the entities that followed Mendel's laws. But their presence had to be inferred from the visible phenotypes using those laws.

Each of the two alleles for a trait is transmitted into an egg or sperm with equal probability. This was Mendel's rule of *independent segregation*. Genes for different traits are transmitted independently of each other: this is Mendel's rule of *independent assortment* that has many exceptions because genes on the same chromosome tend to be inherited together. Whether a trait follows Mendel's rules can be inferred from inspecting family trees (or pedigrees) to see whether there is a precise fit between Mendelian predictions and the presence of a trait in individuals in a pedigree. Such inferences are subtle and typically require a lot of data. The Eugenics Records Office set out to collect that data on a massive scale.

Though, in practice, Davenport used persuasion rather than force to collect these data, he was adamant that the public good demanded their collection at any cost. The passage with which this chapter began continues:

> Society has not only the right . . . but the duty, to make a thorough study of all of the families in the state and to know their good and bad traits. It may and should locate traits of especial value such as clear-headedness, grasp of details, insight into intricate matters, organizing ability, manual dexterity, inventiveness,

> mechanical ability and artistic ability. It may and should locate antisocial traits such as feeble-mindedness, epilepsy, delusions, melancholia, mental deterioration, craving for narcotics, lack of moral sense and self-control, tendency to wander, to steal, to assault and to commit wanton cruelties upon children and animals. It may and should locate strains with an inherent tendency to certain diseases such as tuberculosis, rickets, cancer, chronic rheumatism, gout, diabetes insipidus, goitre, leuchemia, chlorosis, hemophilia, eye and ear defects and the scores of other diseases that have an hereditary factor. It should know where the traits are, how they are being reproduced, and how to eliminate them. It should locate in each country the centers of feeble-mindedness and crime and know what each hovel is bringing forth.

By the time that the Eugenics Records Office had started its data collection, Davenport was already convinced that all these traits were determined by genes. The reason for this was that, as the years went by, he showed less and less concern for the subtlety of Mendelian predictions for the distribution of traits in a pedigree. For him, clustering of a trait in any pedigree was sufficient for genetic inference and eugenic intervention. It came to be the preferred methodology of eugenicists. They found a signature of genetic causation wherever it was convenient.

IQ AND INVOLUNTARY STERLILIZATION

From the perspective of promoting eugenics in the United States, Davenport's new institute was founded at an opportune time. In 1908, Henry H. Goddard, whose official position was Director of Research at the Vineland Training School for Feeble-Minded Girls and Boys in Vineland, New Jersey, had imported the Binet intelligence trait from France to the United States and had begun administering it to thousands of individuals. Though Binet had intended his test only as a diagnostic tool to identify children who needed extra help for any reason, including socioeconomic background, Goddard willfully misinterpreted the tests to be measuring innate intelligence.

Goddard's data supposedly showed that test results mapped seamlessly with independent judgments of intelligence made by professionals who had been working with the supposedly feebleminded. (We will return to the methodology of these tests in the seventh chapter to show how these results, even if reported accurately, were an artifact of the design of these tests.) These tests were soon modified to meet the US context better and, thanks to Lewis Terman of Stanford University, came to be called IQ (for intelligence quotient) tests. During World War I, these tests were used to determine the optimal roles for army recruits, whether they deserved to be foot-soldiers or officers; by the end of the war, over one million seven hundred thousand recruits had been tested generating the largest data set on cognitive functioning the world had ever seen.

IQ enthusiasts insisted that these tests measured basic cognitive function. Skeptics, including many officers of the US Army pointed out that test scores depended critically on recruits' cultural and economic background. By and large, all such criticism was ignored because the test results pandered to the prejudices of the day. Test results were supposed to remain constant throughout an individual's life (though, at the time, there could be no data that supported any such claim: there simply had not been enough time since the tests were invented). They were supposed to vary systematically across races. At the time, "race" designated what we now typically consider as ethnic groups rather than groups identified solely by their skin color. Thus, the tests helpfully showed that Northern European groups (the so-called Nordic race) had superior IQ than those from Southern and Eastern Europe, as well as Jews, let alone people of color. Thus, immigration of inferior groups was supposed to be resulting in the degeneration of the nation, an effervescence of one of the most resilient racist tropes of US politics.

Finally, IQ enthusiasts insisted that IQ was inherited through the genes. Thus, they validated eugenic objections to feeblemindedness. Once again, data in support of any such claim were sketchy, to put it kindly, and based on biased studies of a few family pedigrees which were supposed to be filled with vagrants, alcoholics, the sexually promiscuous, the feebleminded, and other undesirables. Much of these data were produced by Davenport's institute. Thus, not surprisingly, almost all IQ enthusiasts were eugenicists. The pressing question for them was what to do about these supposedly genetically inferior people, especially their continued influx into the country.

Eugenicists organized themselves to play an increasingly strident role in US public policy. Davenport's minions were prominent in those efforts. Eugenicists throughout the United States pursued a two-pronged policy intervention. For those deemed unfit who were already in the country, they demanded sterilization. For would-be immigrants, they demanded exclusion. Thus, eugenics came to pervade public policy on immigration and involuntary sterilization for decades beginning in the 1910s. A central figure was Harry Hamilton Laughlin, a Missouri school teacher whom Davenport had recruited to the cause and who became the first (and, as it turned out, the only) Superintendent of the Eugenics Records Office.

Eugenicists did not initiate attempts to restrict immigration to prevent the entry of perceived "undesirables" into thc country. These moves dated back at least to 1875; moreover, the Chinese Exclusion Acts of 1882 and 1902 as well as the so-called Gentlemen's Agreement of 1907–1908 with Japan all severely restricted Asian immigration though with fewer restrictions for the Japanese compared to other nationalities. Other Acts followed culminating in the Immigration Act of 1924 that curtailed immigration from eastern and southern Europe. This Act followed a report prepared by Laughlin that claimed, on the basis of fudged data, that recent immigrants were biologically inferior to native-born US citizens. Eugenicists hailed the "biological wisdom" of this law; it remained in force for a half-century. It was not substantively amended until 1968 when, during the presidency of Lyndon B. Johnson, immigration quotas were set on the basis of the size of the population of the country of origin.

In the United States, sterilization laws also preceded the eugenicists. In the late nineteenth century, in many states, "retarded" women were institutionalized during their reproductive years. State laws were passed to prevent marriage by alcoholics, epileptics, the retarded, and persons with chronic disease. Several states considered the castration of criminals as a remedy for such "undesirables," and a few superintendents of asylums did carry out mass castrations. But these measures did not enjoy wide popularity until the 1899 advent of a new technology: vasectomy, promoted by a young physician, Albert Ochsner, who eventually became Professor of Surgery at the University of Illinois at Chicago. Ochsner was also the first to recognize the eugenic possibilities of vasectomy because it did not impair sexual performance while preventing procreation.

Almost immediately, the cause of vasectomy was taken up in earnest by Harry C. Sharp, a surgeon at the Indiana Reformatory in Jeffersonville. In 1907, Indiana became the first state to enact a law enabling compulsory sterilization and Sharp and his successors went on to perform 450 vasectomies on incarcerated men, not always with legal sanction. Between 1907 and 1913, the legislatures of sixteen states passed sterilization bills; of these twelve went on to become laws while the other four were vetoed. However, between 1912 and 1921, eight of these laws were challenged and seven overturned by courts; Washington state was the sole exception.

A second wave of eugenic sterilization laws emerged in the 1920s with strong support from Laughlin who had become their foremost scientific advocate even as his status as a competent scientist gradually dissipated during a period in which genetics matured as a discipline in the United States. In the early 1920s, frustrated by the reversals they had experienced in the courts, eugenicists organized themselves and coordinated efforts to take the issue to the Supreme Court. Laughlin, in consultation with legal scholars, drafted a model sterilization law that was supposed to be able to survive legal challenges. In Virginia, eugenicists duly enacted a sterilization law based on this model in 1924 that they wanted to be tested in the courts. Their reasoning was that, if it survived, it would serve as a precedent for other states.

The test case was a seventeen-year-old girl who was supposed to be a "moral imbecile." Carrie Buck, who had already borne an illegitimate child, Vivian, had been committed to the Virginia Colony for Epileptics and Feebleminded where her mother, Emma Buck, also certified as being feebleminded, had already been in residence since 1920. The Colony's board of directors ordered Carrie Buck to be sterilized. A court-appointed guardian challenged the order and the case was set to determine the legality of sterilization in Virginia and, so the eugenicists hoped, for the entire United States.

Carrie Buck was given an IQ test which supposedly showed her mental age to be nine. Thus, Emma and Carrie Buck provided two generations of feebleminded. However, making the case for sterilization watertight required three generations to be affected. It was left to see what Vivian could contribute. She was seven months old. A Red Cross worker was prevailed upon to claim that little Vivian had a "look" that was "not quite normal." An official from the Eugenics Records Office gave Vivian an IQ test that supposedly showed, *at seven months*, she was below average for a girl her age. Meanwhile, Laughlin examined the pedigree and, without once having examined any of the individuals, provided testimony that Carrie Buck suffered from hereditary feeblemindedness. The first court and a subsequent appeals court in Virginia both ruled against Carrie Buck. Finally, in 1925, the case made it up to the Supreme Court as *Buck vs. Bell*; Bell happened to be the name of the person who had recently become the Superintendent of the Colony.

That the case got to the Supreme Court was just what the eugenicists had hoped. The court ruled eight to one in favor of compulsory sterilization. The majority opinion was written by Oliver Wendell Holmes, Jr., widely regarded as one of the century's greatest US jurists. Drawing a connection between patriotism and eugenics, Holmes argued that the principle that allowed society to require mandatory vaccination was sufficient to embrace compulsory sterilization. With respect to the Bucks, he famously concluded: "Three generations of imbeciles is enough." Posterity does not regard it as one of Holmes' finer moments.

Shortly afterward, Carrie Buck was sterilized. Vivian only lived long enough to go through second grade before she died of an intestinal disorder in 1932. Her teachers had uniformly regarded her as very bright.

GENETICS AGAINST EUGENICS

The historian, Mark Largent, has carefully collected the statistics of involuntary sterilization in the United States. Thirty-two states had laws permitting this practice; another five practiced it anyway. Eugenics was not the original or only motivation for involuntary sterilization in the United States. In the nineteenth century, many proponents cited immediate "benefits" of male sterilization through castration including changed behavior (for instance, decreased masturbation which was viewed as a health problem) and reduction of the costs of incarceration of the unfit; others defended punitive castration of rapists and sex offenders (such as child molesters) as well as homosexuals and others perceived to be sexual deviants. However, starting in the early twentieth century, proponents of sterilization were largely motivated by eugenic arguments. They were incredibly successful.

The practice continued for eight decades. The last recorded legal involuntary sterilization in the United States took place in Oregon in 1981. By then, over sixty-four thousand people had been sterilized in the country; California alone contributed more than twenty thousand. Involuntary sterilization of prisoners continued in California into the twenty-first century with 148 cases between 2006 and 2010. Nevada and New Jersey were somewhat unique in having compulsory sterilization laws, going back to 1911 in both states, but they had no recorded case of eugenic sterilization. Enthusiasm for sterilization was highest in the 1930s and declined after that as competent biologists increasingly began to reject any sound genetic basis for such eugenic measures.

These biologists included some of the most prominent geneticists in the United States. Even those who were sympathetic to some eugenic measures, including voluntary sterilization of people with severely deleterious genes, were opposed to involuntary sterilization either on grounds of respect for personal liberty or because they recognized the extent to which eugenic claims were biologically unsound. For instance, contrary to eugenic claims, there was no credible evidence that feeblemindedness was inherited. In the 1920s, Herbert Spencer Jennings, a prominent zoologist at Johns Hopkins University, took Laughlin to task for the shoddy statistics in his immigration studies. Jennings also pointed out that sterilization of the feebleminded was ineffective and irrelevant. Raymond Pearl, also at Johns Hopkins, and no enemy of eugenics were it done right by his lights, dismissed the analysis of pedigrees on which sterilization advocates relied as "old-fashioned rubbish."

Thomas Hunt Morgan at Columbia University had become the greatest geneticist of his generation by using the fruit fly, *Drosophila melanogaster*, to show how genes are linearly arranged on chromosomes and then mapping hundreds of phenotypic traits to genes at precise locations (or *loci*) on these chromosomes. In the 1920s, he explicitly and publicly repudiated the claims of eugenicists such as Davenport and Laughlin pointing out that they had ignored the complexity of the interactions between genes and their environment during development from a fertilized egg to an adult organism.

One of Morgan's students was Hermann J. Muller who later rose to prominence at the University of Texas by showing how *X*-rays can cause mutations on chromosomes. Muller was an ardent eugenicist and communist who left Texas for the Soviet Union with the hope of converting Stalin to the cause of eugenics. Yet, even he rejected the pedigree work on which involuntary sterilization decisions were being based in the United States. He said so, publicly, at the Third International Congress of Eugenics in New York City in 1932. Thus, at the same time that eugenic legislation requiring involuntary sterilization was emerging triumphant as public policy in the United States, eugenics was in intellectual retreat among competent biologists.

Across the Atlantic, in Britain in the late 1930s, the same eugenic policies came under scathing attack from J. B. S. Haldane, a mathematical population geneticist and one of the founders of modern evolutionary theory. Haldane was yet another biologist who had once been somewhat sympathetic to eugenics before encountering the involuntary sterilization policies in the United States—and the measures they spawned in Germany (on which there will be more shortly)—that drove him into the opposing camp. In lectures delivered at Birmingham University in 1937, and published as *Heredity and Politics* in 1938, he brought his stinging wit to bear on the subject and provided what was at that time the most incisive biological critique of eugenic measures. Haldane was a mathematical geneticist and used calculation after calculation to show how ineffective sterilization would be. In particular, most disease-related genes are recessive, which means that only those who are homozygous could be identified through their phenotypes. Even if all those who had the undesirable trait were sterilized, the gene would persist in heterozygotes who could potentially have homozygous children when they mated with each other. (If two asymptomatic heterozygotes mated, on the average, a fourth of their children would be homozygotes.) Sterilization would achieve very little at great costs of loss of personal liberty.

Haldane was blunt in his dismissal of eugenic sterilization. In one engaging example, he quoted data about children in rural schools in one English county:

> Of sixty-three children with an intelligence quotient below 80 percent, no less than twenty-five were the offspring of parents born in the same village. On the other hand, of thirty children awarded free places in secondary schools [for superior academic performance], only two were the children of parents born in the same village.
>
> It might be thought that the parents of the backward children were the village idiots who had never had a chance of leaving their houses. This was not the case. One had been a soldier, another was a carrier, and so on. It is at least

> arguable that the backwardness of the children was largely due to inbreeding, which presumably caused recessive genes [affecting intelligence] to appear in the homozygous condition.

In response, with British understatement, Haldane proposed a novel eugenic measure: "it is likely that the introduction of motor omnibuses into our rural areas will prove to be a eugenic measure quite as valuable as sterilization." The reasoning is simple: inbreeding becomes less likely as people move around. They meet potential partners elsewhere. (Haldane's model predicts that there would be a negative correlation between the prevalence of recessive genetic diseases and the frequency of bus services between British villages in the 1930s. To the best of my knowledge, this prediction has never been tested.)

On eugenic sterilization in the United States, Haldane was slightly less understated: "I personally regard compulsory sterilization as a piece of crude Americanism like complete prohibition of alcoholic beverages." (Haldane was referring to Prohibition in the United States in the 1920s, a measure as absurd as eugenic sterilization from the perspective of individual freedoms though obviously with less harmful other consequences.) Pointing out that the propagation of disease genes could be discouraged by a host of less intrusive measures that did not violate individual freedoms, Haldane went on: "It is perhaps characteristic that in the United States sterilization is legal while contraception is of very doubtful legality." For many geneticists, Haldane's book served as a clarion call to the task of refuting the "pseudo-scientific" assertions of Laughlin and others of his ilk.

EXPORTING STERLIZATION

Eugenic enthusiasm in the United States was probably the main reason why, in 1928, Alberta became the first Canadian province to adopt an involuntary sterilization law very similar to those found south of the border. Over the next sixty years, officials across Canada legally sterilized almost three thousand people without their consent. A class action suit brought by the victims in the 1990s resulted in an award of C$ 142 million in damages to nine hundred and fifty of them.

However, the strongest influence of US eugenic sterilization programs was felt across the Atlantic, in Germany. In 1924, an Austrian-born thirty-five-year-old former corporal of the German army, who was incarcerated in a Munich prison for leading a mob revolt against the state, was spending his time absorbing eugenics texts including a textbook on "racial hygiene" by three prominent German biologists, Erwin Bauer, Eugen Fischer, and Fritz Lenz, first published in 1921. Many of these texts expounded the doctrines of Davenport and other US eugenicists such as Paul Popenoe, an agricultural researcher who had coauthored a popular textbook on eugenics in 1918 and was a strong proponent of involuntary sterilization of those deemed unfit.

The imprisoned ex-corporal also began to follow the views of Leon Whitney, at that time the president of the American Eugenics Society. He was also full of admiration for Madison Grant, a prominent progressive whose achievements included co-founding the New York Zoological Society in 1894 (better known as the Bronx Zoo). Grant was even more famous for a racist diatribe from 1916, *The Passing of the Great Race*. It bemoaned the corruption of the Nordic race in the United States by the presence of Jews, Negroes, Slavs, and other undesirable races.

For the young Austrian prisoner, Grant's book became "his Bible." In the early 1930s, he sent adoring letters to Whitney and Grant acknowledging their influence. By then the writer was no longer a prisoner but a rising star in German politics. His name was Adolf Hitler. While in prison he had written the first volume of his manifesto, *Mein Kampf*, which drew extensively from Grant's diatribe. What had impressed Hitler most about the United States were the sterilization laws that had spread across most of the country. All the way to the Nuremberg trials in 1946 the Nazis would remind the world with some justice that, at least with respect to eugenics, they were only following US precedents.

In 1934, the new Nazi government of Germany enacted a "racial hygeine" law largely based on the model law that Laughlin had drafted. The Nazis appreciated Laughlin's efforts. Later that year, he was rewarded with an honorary degree from the University of Heidelberg for his work. By the end of that year, Hereditary Health Courts that had been set up in Germany had approved more than sixty-four thousand sterilizations. The German Supreme Court ruled that the "racial hygeine" law also applied not only to Germans but even to non-Germans living in Germany who thus also became subject to sterilization.

According to Philip Reilly, who has written extensively on the history of sterilization, German practice did not distinguish between men and women; both were equally targeted for involuntary sterilization. Reilly notes that historians "have estimated that under the powers granted to the Hereditary Health Courts, between 1934 and 1944 (when the population was 73 million) German doctors sterilized at least 400,000 persons, including the mentally ill, the mentally disabled, the deaf, persons with tuberculosis, homosexuals, gypsies, and, of course, Jews." This figure may well be a radical underestimate. According to Reilly, the exact number of involuntary sterilizations in Nazi Germany will never be known.

In 1945, the outside world first became aware of the Nazi death camps and the mass exterminations that had been carried out in them. In much of

Western Europe and in North America, besides of course in Israel, the horror of this period of European history has never fully dissipated. In 1946, during the Nuremberg trials, the outside world was also first told of the Nazi eugenics program including its sterilization policies. The word "eugenics" has never quite meant the same since then. While the Nazi implementation of eugenic sterilization remains horrifying, it should be acknowledged that they were not entirely inaccurate in claiming that they were guided by US practice which they had perfected. While eugenics may not be of US vintage, involuntary sterilization, sometimes on a large scale, certainly is.

There were a few exceptions to this general post–World War II rejection of eugenics as state policy. Later in this book, we will encounter odd cases such as the city-state of Singapore where official eugenic policies have continued to be promoted—though, admittedly, in no way similar to Nazi atrocities. But, especially in the global North, "eugenics" became a dirty word for a generation until attempts at its rehabilitation began emerging in the 1980s when new developments in molecular genetics spawned a brave new program of liberal eugenics supposedly based on individual freedom and the rights of parents to make choices for their possible children. Parents, according to liberal geneticists, should have the right to decide what traits their children should have. Gene editing technologies prior to CRISPR provided the basis for liberal eugenic hopes; CRISPR-based methods have made the prospect immediately tangible. Without sterilization. Even without abortion. And certainly without elimination in death camps.

THE WATSON SCANDAL

When James D. Watson of DNA double helix fame became director of Cold Spring Harbor Laboratory in 1968, he discarded much of the eugenics literature that had accumulated in its library going back to the Davenport era. In the late 1960s, Watson was following the norms of the day: the horrors perpetrated by the Nazis in the name of eugenics seemed to leave little other choice. (Scholars have since noted that some of this material could have been of great historical interest. Watson should have consulted some historians. But, as we shall see, not doing so is one of his lesser sins.)

There is some irony in Watson's discarding of Davenport's material. Among molecular biologists today, Watson is probably closer in spirit to Davenport than any other figure. *Davenport's Dream* republishes an article by Watson, "Genes and politics," that originally appeared in the 1996 Cold Spring Harbor Annual Report. In it, Watson not only endorsed human germline interventions to remove genes for disease but was even willing to tolerate such intervention for genetic enhancement: "we will someday have gene therapy procedures that will let scientists enrich the genetic makeups of our descendants." The only potential eugenic programs that he rejected were those that would be controlled by governments. Liberal eugenics was just fine.

Like Davenport, Watson was convinced of the primacy of genes as causes of human behavior. He denigrated what he called the "'not in our genes' politically correct outlook of many left-wing academics." *Not in Our Genes* was the title of a well-known book by Dick Lewontin, Steve Rose, and Leon Kamin. Watson's attack was gratuitous; while dismissing them as "left wing," he had no response to their arguments against genetic determinism. Another group that drew Watson's ire was Science for the People, based in Cambridge, Massachusetts. Biologists such as Jon Beckwith and Jim Shapiro, who were members of that group, were castigated by Watson for advocating public scrutiny of the new recombinant DNA technology of the 1980s followed by social regulation of its use.

Much of Watson's article was spent promoting the Human Genome Project (HGP) that he had headed in its initial years. He correctly claimed credit for funding the Ethical, Legal, and Social Implications (ELSI) studies of the human genome as part of the Project. However, he noted, that he created this program so that no one could accuse him of perpetuating the eugenic legacy of the Cold Spring Harbor Laboratory through the HGP. In the context of a discussion of Nazi eugenic programs, he denounced "pseudoscientific theories of race superiority and purity." For once, breaking with Davenport even in spirit, he railed against the use of IQ data during and after World War I "to justify the discriminatory segregation laws that effectively made America's black population second-class citizens."

However, at some point after 1996, Watson changed his mind about race and intelligence. In 2007, he made headlines by telling a British journalist that he was "inherently gloomy about the prospect of Africa [because] all our social policies are based on the fact that their intelligence is the same as ours, whereas all the testing says, not really." When it came to a presumed equality of intelligence between blacks and whites, he added "people who have to deal with black employees find this not true." The interview had consequences. Though there was no public evidence that he had ever acted on such racist views, he was removed from his position as chancellor of the Cold Spring Harbor Laboratory. He was, however, allowed to retain an office on campus. *Davenport's Dream*, published a year later, reprinted the paper discussed earlier which was no longer consistent with what Watson now believed about race and intelligence.

The 2007 interview was not a temporary aberration. In 2019, Watson reaffirmed his views on race and intelligence even more firmly. Asked whether he wanted to retract his earlier remarks, he replied: "No, not at all. I would like for them to have changed, that there be new knowledge that says that your nurture is much more important than nature. But I haven't seen any [such] knowledge. And there's a difference on the average between blacks and whites on IQ tests. I would say the difference is, it's genetic." It is perhaps only fitting that Watson should have spent the bulk of his career directing an institution associated with US eugenics during its most grotesque phase. In Watson, at least, Davenport's dream lives on.

ESTABLISHING HUMAN GENETICS

It would be misleading to suggest that revulsion against the Nazis alone led to the general disintegration of eugenics programs after World War II. While the Nazi atrocities had made defense of eugenics awkward, perhaps even unacceptable in polite society at least in North America and Europe, there is much more to the story of the decline of eugenics. Starting in the 1930s, a discipline of *human genetics* began emerging and it was distinct from eugenics in the sense of being overtly concerned with establishing scientific results rather than influencing public policy. One important milestone was the publication in 1936 of the first map of genes on a human chromosome by Haldane. He placed six loci on the *X* chromosome and calculated distances between them to establish the following order: achromatopsia (total color blindness); xeroderma pigmentosum (severe light sensitivity of the skin); Oguchi's disease (night blindness with golden retinal pigmentation); epidermolysis bullosa dystrophica (a skin disease); retinitis pigmentosa with deafness; and retinitis pigmentosa without deafness. All but the last of these were recessive meaning that heterozygous women could be carriers who could transmit a disease to their sons without suffering from it themselves. On the average, half of the sons of such a woman would suffer from the disease. This half consisted of those who inherited an *X* chromosome with the disease allele; the other half would inherit the mother's other *X* chromosome.

Shortly afterward, Haldane and Julia Bell showed how genes for hemophilia and color blindness were linked on the *X* chromosome. Both of these genes are also recessive. Haldane then published the wonderfully titled "Blood royal: A study of hemophilia in the royal families of Europe" as his next foray into human genetics. He drew a pedigree showing the prevalence of hemophilia in the princes of Europe. Because of royal inbreeding, in spite of being caused by a recessive gene, this disease appeared far more frequently in the royal pedigree than it would in a similar pedigree drawn from their subjects, that is, the general public which does not inbreed quite as assiduously.

A careful analysis of the pedigree showed that the source of the hemophilia gene was Queen Victoria's father. Haldane explained: "The gene must have originated by mutation, and the most probable place and time where the mutation may have occurred was in the nucleus of one of the testicles of Edward, Duke of Kent, in the year 1818." Haldane's eugenic message was obvious: to prevent hemophilia would have required the sterilization of Europe's royalty. But the most telling part of Haldane's analysis was that it showed the real power of human genetics at its sharpest: it could tell not only where but *when* European royalty had acquired a mutation for hemophilia and had begun spreading it among themselves through systematic inbreeding. Royalty could reduce the risk of hemophilia, and its burden as a disease, by marrying commoners. (Or, of course, by refraining from breeding.)

The message was also political in another way. This was the period when many scientists including Haldane were drifting into the Communist Party because the traditionally dominant political parties of Europe were indulging in appeasing Hitler. Haldane was pointing out what Nazi eugenic principles implied for European royalty. He was also challenging the mystique, the aura, that many in Europe, and particularly Britain, still associated with royalty. His message was political and intended to have a broad public appeal. "Blood royal" appeared in a nonacademic journal, *The Modern Quarterly*, which had just been founded and was expected to have a wide audience. In retrospect, we can view this analysis as one of the most important early steps in liberating human genetics from eugenic shackles. It was particularly important because Haldane was both one of the most prominent evolutionary biologists of his time and a public figure known for popular scientific writings and pithy opinions. At the time when "Blood royal" and also his book, *Heredity and Politics*, were published, he was a public figure and probably the most culturally influential geneticist in the English-speaking world. We have met him earlier and will encounter him again in our discussions of eugenics.

Haldane was also not alone. In the mid-1930s, as we saw earlier, Hermann J. Muller, who never abandoned his eugenic ideals, was also denouncing what eugenics had become, especially in the hands of Laughlin and other followers of Davenport who had been advocating involuntary sterilization. As a movement, eugenics was in decline: in 1932, the Third International Congress of Eugenics in New York City attracted less than a hundred participants. There was no fourth congress. With the knowledge that comes from hindsight it seems clear that, by the 1930s, the advocacy of involuntary sterilization had doomed any sympathy for traditional eugenics among competent geneticists.

The Nazi atrocities made the rejection of eugenics move forward more rapidly and more comprehensively after 1945. In Britain, one figure stands out: Lionel Penrose. In the 1930s, Penrose had shown the complexity of the genetics of mental dysfunction. His pathbreaking study, known as the

Colchester survey, was published in 1938: it showed beyond credible controversy how worthless had been the early twentieth-century "genetic" work on feeblemindedness. Ironically, Penrose was appointed *Galton* Professor of *Eugenics* in 1945 at the University College London (where Haldane already was Professor of Biometry). In his Inaugural Address in 1946, Penrose dismissed involuntary eugenic sterilization. "Only a lunatic would advocate such a procedure," declared the new professor of Eugenics while discussing procedures for eliminating deleterious genes. For decades, Penrose worked to have his job description changed. He finally succeeded in 1963, when he became professor of Human Heredity shortly before retirement in 1965. By that time, Penrose had used his prestige and authority to systematically remove references to eugenics within mainstream human and medical genetics.

DAVENPORT'S DREAMS TODAY

This leaves us with a glaring question: What was Cold Spring Harbor Laboratory thinking when it published *Davenport's Dream* in 2008 containing within it a facsimile of *Heredity in Relation to Eugenics*? The editors were well aware that they were moving into suspect territory. They note:

> It is unusual to reissue a book that has long been unavailable, with the content being outdated. Charles Davenport's *Heredity in Relation to Eugenics* was published almost 100 years ago, and its subject matter—eugenic studies in the early twentieth century—was consigned in the 1940s to what is now referred to as "pathological science." Why, then, revisit it?

They acknowledge the book's historic interest which is perhaps by itself a fairly good reason to keep it in print. But that is not what had motivated republication. Instead, in their words:

> the most compelling reason for bringing Davenport's book once again to public attention is our observation that although the eugenic plan of action advocated by Davenport and many of his contemporaries has long been rejected, the problems that they sought to ameliorate and the moral and ethical choices highlighted by the eugenics movement remain a source of public interest and cautious scientific inquiry, fueled in recent years by the sequencing of the human genome and the consequent revitalization of human genetics.

Matt Ridley, the conservative journalist and popular science writer, adds in his Foreword to the volume: "Charles Davenport had the best of intentions." I cannot help asking: *Really*? Have you read him?

Eugenics is back. *Davenport's Dream* includes a wide array of pieces by prominent figures commenting on *Heredity in Relation to Eugenics*. Most, though not all, of these pieces are starkly revisionist, attempting a new assessment depicting Davenport and his project in a better light than how he was presented in earlier writings, especially by historians of science. The revisionist agenda underscores the extent to which eugenics has come back with a vengeance. It is back because the principals of the HGP accurately foresaw the eugenic potential of increasingly detailed knowledge of the human genome. To Watson's credit, as the first director of the HGP, he made the ELSI program a core component of the project. There is no evidence to suggest that Watson's racist beliefs, even if they had already been simmering at the back of his mind, played any role in what the ELSI program nurtured. Work supported by the ELSI program routinely raised the specter of eugenics to question the ethics of pursuing the HGP. (For the sake of full disclosure I should add that I was one of those generously funded by the ELSI program during its earliest phase even though I was a vocal critic of the HGP. We will encounter some of those criticisms in the third chapter.)

At the time the HGP was initiated, prenatal screening of embryos for a suite of genetic diseases was already available. Thus, through selective abortion of embryos with these deleterious genes, eugenic intervention was possible and, to a considerable extent, practiced without ever using the word. More importantly, there had already been several rudimentary and generally unsuccessful attempts at gene therapy in somatic cells by introducing functional versions of deleterious genes. Advocates of the HGP promised that new technological advances spawned by the project would make gene therapy—and gene editing in general—medically viable. However, it was more than a decade after the completion of the HGP that the advent of CRISPR technology began to fulfill that promise. In 2008, when *Davenport's Dream* was published, CRISPR technology was yet to come and eugenics still remained a somewhat abstract possibility; post-CRISPR it has emerged as a tangible choice already available to tantalize the present generation of would-be parents.

What is more important is that, in spite of all the revulsion generated by what was done in its name, eugenics has been able to stage its post-HGP comeback because the basic premises of eugenics have never been convincingly ethically or scientifically fully refuted. Yes, we may abhor involuntary sterilization and murder, as we should. But these are not the only available eugenic measures. For instance, in Britain, where involuntary sterilization was never practiced, few eugenicists had sympathy for involuntary sterilization, let alone for death camps for the unfit. For many British eugenicists, the preferred policy was to offer financial incentives for the supposedly genetically superior to reproduce more frequently while similarly discouraging the supposed inferior from doing so. Even in the United States, in the early part of the twentieth century, women's rights activists such as Margaret Sanger were ardent eugenicists; for them, eugenics provided a rationale for birth control, their preferred eugenic strategy, though some of them including Sanger herself were also willing to embrace involuntary sterilization for extreme cases of feeblemindedness.

Does ethics rule out eugenics? Presumably, no one will seriously dispute that it would be a good thing if incurable debilitating genetic diseases such as Huntington's disease or myotonic dystrophy were to be permanently eliminated from the human population. Now, if someone were to propose eliminating such diseases by intervening only in somatic cells, then there would presumably be no occasion for controversy. Eugenics creates controversy because, for one reason, it has been associated with policies such as sterilization that have viciously trampled on the rights and liberties of individuals and, for another reason, it proposes to interfere with the germline of people, what is sometimes called the human gene pool. But, as we just noted, these two aspects of traditional eugenic policies can be separated from each other. What is fundamental to eugenics is conscious interference in the human germline, not coercion. As we shall routinely see in this book, this prospect of altering the human gene pool generates disquiet in a wide variety of people independent of the issue of coercion. Nevertheless, compelling arguments to ban all interventions into human germlines have proven hard to formulate.

In recent decades, a small but vocal group of liberal eugenicists have emerged advocating germline modification of human embryos guided by parental choice with no role for social coercion. However, these liberal eugenicists go beyond germline editing of disease genes which is ethically the easy case. They advocate genetic enhancement, that is, germline intervention to augment qualities of body and mind that are considered desirable. The ethical questions now become much more controversial. Do we want children with better mathematical ability? Or better baseball skills? Or lighter skins? Should we? Who decides? But, to discuss these questions now would be to get ahead of the story of this book. In spite of much discussion, there is no consensus on how we should answer these questions. So, it seems quite appropriate that the editors of *Davenport's Dream* chose to help reopen discussions of the desirability of eugenics. But it still remains doubtful that Davenport's personal example has anything positive to contribute. The editors may have served society better if they had noted the horrors enabled by Davenport and his minions. Perhaps a more worthwhile endeavor would have been entitled "*In Spite of Davenport*."

Is eugenics scientifically and technologically feasible? When an emerging generation of human geneticists liberated themselves from eugenics in the 1930s (and we should think of figures such as Haldane and Penrose), one of their favorite stratagems was to point out that the policies supported by eugenicists such as involuntary sterilization were scientifically absurd: they could not achieve their intended goals. Recessive genes are phenotypically undetectable in heterozygous people (those with only on copy of the gene); as Haldane pointed out, involuntary sterilization programs would be ineffective at eliminating these genes. Haldane also pointed out that these programs were an affront to individual liberties and, thus, the ethical price that would have to be paid was unacceptable compared to the very modest benefits achieved. These scientific arguments against eugenics were politically the most influential ones in the 1930s when the eugenic project had begun its generation-long decline.

Since then, at least on the surface, these scientific problems have been resolved for some simple cases such as diseases caused by dominant single genes. Genome sequencing has become straightforward—and cheap—thanks to the technologies spawned by the HGP. Detecting a heterozygous carrier of a gene is technologically trivial. More importantly, screening embryos for the presence of individual genes is equally straightforward. But what has truly changed the game is the emergence of reliable, accurate, and cheap gene editing through CRISPR. We can edit an embryo's genes without doing it any harm, let alone killing it.

However, once we probe under the surface, the technological situation becomes much less promising. The list of diseases that can be eliminated by intervening in a tractable number of genes, even with CRISPR, is small. Right now, the list of traits that are generally held to be worth promoting and that can be enhanced by intervening in a tractable number of genes is glaringly empty. Organisms are not lumbering robots controlled by their genes; humans less so than all other species. We are constructed during embryonic development using genes as critically important resources but along with an array of other materials needed for construction. These material resources must be available to the developing embryo—and, later, the fetus—in the right way at the right time. Merely editing genes, as we shall see over and over again, can lead to very odd results. What genes end up doing depends on context.

One contribution to *Davenport's Dream*, perhaps the most profound piece in the book, recognizes this problem. Daniel Weinberger and David Goldman, both psychiatric geneticists, emphasize: "the multivalent nature of genetic variation altering human behavior makes the goal of behavioral improvement by eliminating deleterious alleles [genes] an illusory goal. Through selection of so-called 'good' alleles, we might well be able to impoverish human behavioral diversity, but it is unlikely that we would thereby improve the mental health of populations." They might as well have said: "Roll over, Davenport."

Even in the case of well-behaved disease genes, how did we end up in this situation when intervening in genes has become the preferred approach to the most recalcitrant genetic diseases? Why don't we treat the diseases when *and if* they manifest themselves? We turn to that story first. The second half of the twentieth century is the story of the emergence of two new sciences that have changed the course of human history. One was computer science. The other was molecular biology to which we will turn next. Not only has it changed how we view every facet of biology, it has had an indelible impact on medicine by recalibrating how drugs and vaccines should be designed. But, beyond that, molecular biology has singularly failed to provide successful medical interventions into genetic diseases from Huntington's disease through myotonic dystrophy and cystic fibrosis to sickle cell disease. Thus, we get a rationale for the project of intervening at the level of the genes with the hope that cellular reactions will humbly follow genetic dictates.

Chapter 2

Molecular Diseases, Elusive Treatments

"It is almost certain that cancers . . . arise because genes concerned with the regulation of cell division are mutated, partly as a consequence of environmental insults, partly because of unavoidable molecular instability, and even sometimes as the consequence of a viral attack on the genome. Yet the realization of the role played by DNA has had absolutely no consequence for either therapy or prevention, although it has resulted in many optimistic press conferences and a considerable budget for the National Cancer Institute. Treatments for cancer remain today what they were before molecular biology was ever thought of: cut it out, burn it out, or poison it."

—Dick Lewontin, "Billions and billions of demons," 1997.

THE MOLECULARIZATION OF BIOLOGY

Lewontin's observations will no doubt come as a surprise to those who have followed the molecular revolution that has transformed biology since the mid-twentieth century. The iconic event occurred in 1953 when Watson and Francis Crick constructed the double helix model of DNA, the linear chain molecule specifying genes and consisting of a sequence of four nucleotide bases (*A*, for adenine, *C* for cytosine, *G* for guanine, and *T* for thymine). The DNA double helix became a symbol of triumphant molecular biology. Before molecular biology, life was typically studied at the level of organisms; knowing what went on within an organism, especially within a cell, was largely a matter of conjecture. For instance, as we saw in the last chapter, the presence of genes was inferred from whether their phenotypes were inherited following Mendel's rules. Indeed, until the 1940s, genes were generally believed to consist of protein molecules rather than DNA. Their chemical properties made no difference to how they were studied in the laboratory. Molecular biology changed all that forever.

Four biologists were the primary architects of the molecular revolution: Crick in Britain, Linus Pauling in the United States, and Jacques Monod and Francois Jacob in France. Pauling paved the way in the 1930s. He was already a celebrated physical chemist who had shown how the quantum mechanics of electrons within atoms explained the rules of chemical bonding including what valency each type of atom has (that is, how many hydrogen atoms it could bond with to form a molecule). In the 1930s, Pauling collaborated with Karl Landsteiner, an Austrian-born researcher at the Rockefeller Institute for Medical Research in New York City (now Rockefeller University), to try to understand the *specificity* of immune reactions, why each antibody, which is a large protein macromolecule, reacted with precisely a single antigen, a fragment of a pathogen that was attempting to invade a body.

Pauling argued that specificity arose because these reactions required a lock-and-key fit between the shape of the antigen molecule and the "active site" on the surface of the antibody. According to this theory, the antigen was captured by the antibody's active site and destroyed. In the 1940s and 1950s, by building tinker-toy structural models of proteins, Pauling and his followers showed that the same principle held for other cases of biological specificity, for instance, how enzymes only reacted with their own substrates and no other molecules. In fact, biological macromolecules such as proteins (which include antibodies and enzymes) and nucleic acids (such as DNA and RNA) all interact through surfaces touching each other with perfect fits of shape. Thus, *structure determines function*. Moreover, because all that is involved is macromolecular surfaces touching each other, living phenomena depend on what, from the perspective of physics, depend on very weak forces. The contrast here is with the chemical bonds that hold molecules together: these require forces that are hundreds of times stronger.

The perspective that Pauling brought to the molecular study of biology was thoroughly *reductionist*: the behavior of wholes, for instance, cells was being explained from the interactions of their parts, that is, the macromolecules that made them up. Moreover, the interactions between the macromolecules were essentially physical; they were present in both animate and inanimate matter. Biology was finally being given a materialist interpretation, a project that goes back to the seventeenth century but had remained a pipe dream until the advent of Pauling's work on molecular biology. *Reductionism*, the idea that this type of explanation will be possible for all phenomena, is a philosophical thesis about the nature of science: we will have more to say about this thesis below and, especially, in the seventh chapter. However, Crick soon threw a spanner into the works of this project, though without realizing what he had done. An ex-physicist, Crick believed himself to be as much of a reductionist as Pauling. Yet, he introduced into molecular biology a type of reasoning entirely foreign to physics. This reasoning involved relying on a concept of *information*.

In the late 1950s, Crick was interested in how genes acted. The DNA double helix model provided some interesting clues. The cylindrical backbone of the model consisted of two helices intertwined around each other. Within this cylinder were pairs of nucleotide bases with one member of each pair attached to each helix. Base pairing followed strict rules: *A* was always paired with *T*, and *C* with *G*. This was called base pair *complementarity*. However, there was no restriction on the sequence of bases along the backbone: it could be anything and the four possibilities at every position allowed for an immense amount of variability. The DNA double helix could thus specify a practically infinite number of different genes with only a few thousand base pairs.

For all its appeal, the helical shape of the DNA molecule has turned out to be biologically irrelevant. What is crucial to its functioning is base pair complementarity. As we shall see, this feature is also crucial for CRISPR-based gene editing: the gene-editing machinery recognizes the targeted gene for editing using complementary base pairs. Complementarity suggested a simple mechanism for the copying of genes during cell duplication: the double helix would unwind and each of the separated strings would serve as a template for bases to attach. Because of base pair complementarity, the

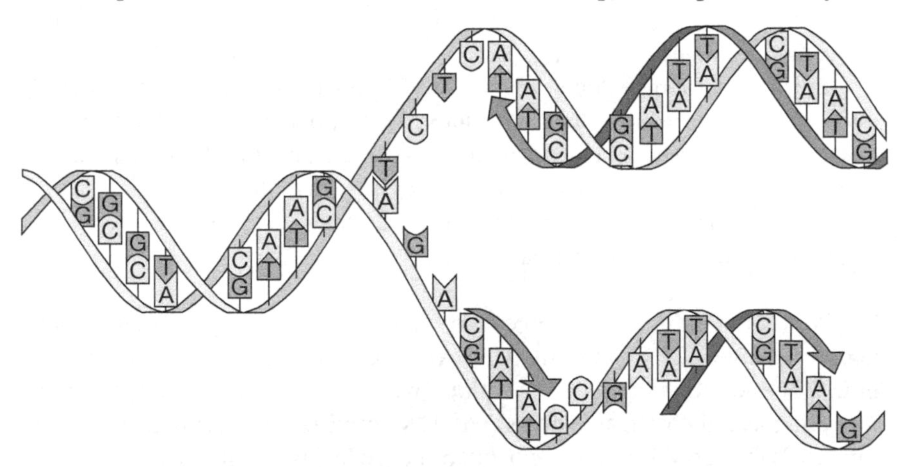

Figure 2.1 DNA Replication. The two helices of the DNA unwind to form a fork. A new double helix forms along each prong of the fork through base pair complementarity Thus, each of the two replicated double helices has one helix of the original DNA.

two double helices that would thus be formed would each be identical to the original. Genes would thus be copied with perfect fidelity. An experiment carried out by Matthew Meselson and Franklin Stahl in 1958 confirmed this mechanism. Figure 2.1 shows this process.

By the late 1950s, it was also clear that the sequence of DNA bases in genes specified the sequence of amino acid residues that formed linear protein molecules. The relationship between DNA and protein came to be viewed as one of *coding* and the postulated genetic code was eventually worked out by the 1960s. It turned out that each amino acid residue of a protein was encoded by three DNA bases and that there was no overlap between these triplets as they specified the sequence of amino acid residues, one by one. The DNA of the gene was first *transcribed* into RNA with a complementary sequence (with one difference: RNA uses uracil [U] instead of T). This RNA was then *translated* into protein at dedicated organelles in the cytoplasm called ribosomes. The whole process constitutes gene *expression*.

In 1958, Crick interpreted these processes as a transfer of *information*: "protein synthesis," he argued, consisted of "the flow of energy, the flow of matter, and the flow of information." What was novel about this formulation was that, whereas the flow of matter and energy was standard physics and chemistry, *information* introduced an entirely novel entity into the mix. Neither physics nor chemistry has any law about information. Strangely, Crick seems to have been blithely unaware of the innovation he had apparently inadvertently introduced. He embedded the flow of information into a core assumption of molecular biology which he called:

> *The Central Dogma*
>
> This states that once "information" has passed into protein it cannot get out again. In more detail, the transfer of information from nucleic acid to nucleic acid, or from nucleic acid to protein may be possible, but transfer from protein to protein, or from protein to nucleic acid is impossible. Information means here the precise determination of sequence, either of bases in the nucleic acid or of amino acid residues in the protein.

The informational interpretation of the most basic genetic processes became central to molecular biology though, over the years, it has come to be challenged on the ground that "information" was no more than a metaphor masquerading as a theoretical concept: Crick's definition was criticized as being hopelessly vague. There was no quantitative measure of information that made any sense. From this perspective, the use of informational concepts in molecular biology was an artifact of the intellectual atmosphere of the 1950s when the first digital computers were being made and odd short-lived disciplines such as cybernetics flourished.

Between Pauling's emphasis on structure and Crick's focus on information, molecular biology emerged in the 1950s with a powerful set of conceptual tools to tackle the problems of life. Two problems were perceived to be of critical importance. The first was the curious goal-directed (or "adaptive") behavior of bacteria which produced an enzyme to digest a sugar only when that sugar was available. For instance, *Escherichia coli* bacteria only produced an enzyme to digest lactose when there was lactose available for absorption from the environment. When lactose becomes unavailable, the bacterial cells turned off production of the enzyme that digested it. This was known as *enzymatic adaptation.* It was supposed to be a "teleological" phenomenon in which a future event (the "goal," that is, the digestion of sugar) directed present behavior (the production of the needed enzyme).

The second was a form of *cooperative behavior* known as the Bohr effect after Christian Bohr, the Danish physiologist who first described it toward the end of the nineteenth century. When blood absorbed oxygen in the lungs using hemoglobin molecules, the presence of a little oxygen in the blood led to more rapid further absorption of the gas until the blood began to become saturated with oxygen and the absorption rate leveled off. Cooperative behavior suggested that the whole was more than the sum of its parts. Somehow the molecular parts collaborated with each other to perform a function better than what they could have done had they not collaborated. Both phenomena, teleological behavior during enzymatic adaptation and cooperative behavior in hemoglobin, were supposed to challenge thc type of physical reductionism that Pauling had championed.

In Paris, Monod and Jacob solved both problems through a mixture of superb experimental work and model building. Enzymatic adaptation was explained by the operon model: the sugar, when present, attached to and removed a molecule from the DNA and allowed the gene for the digestive enzyme to be expressed. Gene expression could also be induced by a molecule other than the sugar if it had the same shape and size. It did not matter at all whether the enzyme could digest this molecule. There was nothing inherently goal-directed about it; in the *lac* operon what was at stake was structure which determined function. The behavior of hemoglobin was explained by their allostery model. Hemoglobin was a macromolecular complex consisting of four chains of amino acid residues held together by an iron-containing heme group of atoms that attached to oxygen and to all four chains. The attachment of the first oxygen atom to the heme group changed the shape of the complex in such a way that more oxygen atoms could fit more easily into the actives site and thus get attached to the complex. Structure again explained behavior in the case of hemoglobin. In the end, reductionism remained unsullied by these challenges. Molecular biology entered the 1960s as a very successful new science.

MOLECULAR DISEASES

Meanwhile, Pauling had turned to the molecular study of genetic diseases. These had been known since 1902 when Archibald Garrod, a London physician, had studied alkaptonuria, a condition in which urine turns black due to the presence of two copies of a mutant recessive gene. Garrod went on to show that there were a number of diseases that could be given a genetic interpretation through the study of pedigrees. These were the ones on which pioneers of human genetics such as Haldane and Penrose focused while they worked to establish the discipline as a distinct discipline independent of eugenics.

In the period immediately after World War II, the genetic disease that came to the forefront of molecular research was also one due to a recessive gene: sickle cell anemia. Pauling, who had begun to be interested in hemoglobin in the mid-1930s, was responsible for drawing attention to this disease within molecular biology. During World War II, Pauling had systematically worked on hemoglobin while researching blood substitutes as part of the war effort. In 1945, a clinician, William B. Castle, from Harvard University drew Pauling's attention to the odd fact that, in sickle cell patients, only blood in the veins (that is, blood lacking oxygen) formed sickle-shaped red blood cells that got stuck in capillaries and caused sickle cell anemia. The role of oxygen suggested that hemoglobin was involved in the distorted shape of the red blood cells and, therefore, the onset of disease.

Pauling, not surprisingly, took up the challenge of elucidating how some physical or chemical difference between sickle cell and normal hemoglobin molecules could account for this observation. He assigned the problem to a graduate student, Harvey A. Itano, as a dissertation project. Itano finally solved the problem using gel electrophoresis, a newly invented technique for detecting electric charge differences between macromolecules. The method used an electrical field set up across a gel between two electrodes. Macromolecules migrate across this field depending on their charge and mass. Itano showed that sickle cell hemoglobin molecules were positively charged relative to normal hemoglobin. Pauling and Itano (and their collaborators) distinguished between having sickle cell disease and the sickle cell trait when patients have a very mild form of sickle cell anemia. The former only had sickle cell hemoglobin in their red blood cells whereas the latter had both sickle cell and normal hemoglobin. In the title of a seminal paper that they published in 1949, the group announced that sickle cell anemia was a "molecular disease."

Pauling and Itano concluded that individuals with the sickle cell trait were heterozygous, possessing genes for both normal and sickle cell hemoglobin. Completely independently of this group, through a series of genetical experiments, the physician, James Neel, came to the same conclusion at about the same time. The normal hemoglobin allele is dominant over the sickle cell allele but dominance is incomplete: that is why heterozygous individuals showed some disease symptoms. Pauling and Itano's work had shown how a mutation in a single gene could alter the physical properties of a protein in such a way as to cause a disease.

It took less than a decade to work out the details of this mutation. In 1956, Vernon Ingram in Britain showed that the difference between normal and sickle cell hemoglobin molecules consisted of the substitution of one amino acid, glutamic acid, by the amino acid, valine. This substitution occurred at a precise position near one end in one pair of the four polypeptide chains in a hemoglobin macromolecule. (Since the two altered chains are identical and specified by the same gene, a single mutation in that gene results in two changes in each macromolecular complex that constitutes a hemoglobin "molecule.") This result began to explain the physics of sickling. Glutamic acid is a charged amino acid residue and, because of this charge, has an affinity for water. In contrast, valine has no charge and is thus more likely to encourage clumping between hemoglobin molecules when it is present. This is why sickle cell hemoglobin formed long fibers and distorted the shape of red blood cells in some situations, especially when oxygen concentration was low. Normal hemoglobin did no such thing.

Thus, in 1956, Pauling felt confident enough to declare: "man is simply a collection of molecules . . . and can be understood in terms of molecules." The declaration presented a stunning molecular vision of life though, after the early 1950s, Pauling contributed little more toward its fulfillment. In the years that followed, he moved away from science into organizing a political crusade for nuclear disarmament. He argued that nuclear bomb testing would cause more mutations and more "molecular diseases"; the most effective strategy for preventing an epidemic of these diseases would be to ban the bombs altogether. The crusade would lead to a second Nobel Prize for Pauling, for Peace in 1962. (He had already been honored with one for Chemistry in 1954.)

But in case we become too enthralled by Pauling's vision and its therapeutic implications, given the rapid development of molecular biology, including the HGP and afterward, we should pause to ask how successfully we treat sickle cell disease today. The answer should be sobering. Some seventy years after Pauling's original work, we have no cure for sickle cell disease, not even a very effective management strategy. Serious anemia is best treated with blood transfusions just as it was in 1949. Much of medical advice given to patients consists of ways to prevent the occurrence of "sickle cell crises" or pain episodes generated by red blood cells being stuck in capillaries. Prevention of crises includes a wide variety of behavioral adjustments including drinking ample water, avoiding high altitudes, and other measures to prevent the oxygen content of blood decreasing too much. Deoxygenated blood increases the chance of sickling; even this rather modest insight was available before Pauling's seminal paper.

One possible conclusion to draw may be that we should give up on molecular medicine at the level of hemoglobin or other protein molecules; rather, if and when we have the technological capacity, we should target the genes themselves. This means we should edit genes as a method of therapy. It should, therefore, not come as a surprise that the sickle cell hemoglobin gene was one of those that the (U.S.) National Academies of Sciences, Engineering, and Medicine prioritized in 2017 for potential editing, at least in somatic cells, and perhaps in the long run even in the germline.

BUT NO MOLECULAR MEDICINE

When it comes to genetic disease and the impotence of molecular medicine, sickle cell disease is not an exception. Take phenylktonuria (PKU), a disease that leads to cognitive disability and has been known since the 1930s to be caused by a single recessive gene. Homozygous persons with two disease alleles cannot digest the amino acid residue phenylalanine which is a standard component of the proteins that we eat. In the 1950s, with contributions from many researchers including Penrose, a diet without phenylalanine was invented for these homozygotes. If they are put on this diet from birth, the disease does not manifest itself.

Starting in the early 1960s, first in the United States and then elsewhere, babies began to be genetically screened for PKU so that they could be immediately put on this diet if they had two copies of the allele for the disease. Sixty years later, this is still what we do today to manage PKU. Even though the prescribed diet is expensive, we have no other option. The gene for this disease has always been rare. Given screening and dietary treatment, the disease has become almost entirely unknown to pediatricians in relatively rich countries except in textbooks touting the success of scientific medicine. While science in the form of genetics and physiology from the pre-molecular era are responsible for this success, molecular biology has made no difference.

Take Tay-Sachs disease, yet another recessive single-gene disorder. If a newborn inherits two copies of the allele for it, then the disease typically begins to appear between the ages of 6–12 months when the baby starts to lose motor abilities such as being able to turn over, sit, or crawl. The disease progressively destroys neurons and most children die by the age of four years. The genetic nature of Tay-Sachs disease was recognized by the 1930s. The malfunctioning enzymes were identified in the late 1960s. In this case, there is no known intervention that prevents the onset of disease and, once again, no semblance of a cure. The only option is screening for the alleles, discouraging reproduction between two heterozygotes, or, if reproduction is initiated, potential destruction of homozygous embryos or fetuses. Once again, molecular biology has so far made no difference.

Indeed, in a thoughtful assessment published in 2000, and after noting the many triumphs of the emergent molecular biology of the 1950s, the historian, Hans-Jörg Rheinberger, observed:

> despite much public praise and hope, [the triumphs of molecular biology] were of quite limited immediate influence on medicine and its practices. In many cases, the results . . . simply did not lend themselves to therapeutic application (as in the case of sickle cell anemia). In other cases they basically sanctioned a practice that was well under way and had developed without the direct impact of molecular biology, as in the case of antibiotics . . . In still other cases, molecular techniques expanded diagnostic potentials, but did not qualitatively change, much less revolutionize the possibilities of metabolic correction.

Rheinberger did not pause to explain why, in the case of sickle cell anemia, even a *full* knowledge of the molecular mechanism did not "lend" itself to therapy despite a half-century of research. Perhaps there is something incomplete about a molecular vision of life? A former molecular biologist, Rheinberger, was not trying to question the medical potential of the discipline. Rather, the piece from which this quotation was drawn was intended to support the potential for a molecular revolution in medicine brought about by the HGP. The crucial ability was supposed to be the capacity to manipulate DNA directly, that is, gene editing.

Indeed, Rheinberger waxed lyrical about the prospects of medicine based on DNA editing:

> The advent of gene technology, genetic engineering or, as some prefer to say, applied molecular genetics, since the beginning of the 1970s has effected a decisive prospective change in the relation between molecular biology and medicine. The emergence of these so-called "recombinant DNA technologies" has created a new situation. With gene technology, the central technical devices of molecular biological intervention have themselves become parts and indeed constituents of the metabolic activities with which, at the same time, they interfere. The scissors and needles by which the genetic information gets tailored and spliced are enzymes. The carriers by which it is transported into the cells are nucleic acid macromolecules. This technique is of potentially unlimited impact.

For all the enthusiasm, notice that this passage was written in 2000, more than fifty years after Pauling's paper on sickle cell disease. Twenty years later, we are still left with promissory notes rather than tangible therapies.

DREAMS OF EDITING GENES

Though success with therapeutic gene editing has only become plausible in the post-CRISPR era, that idea goes back to the 1960s. Two developments in the 1950s paved the way. Both involved viruses. The first was the unexpected observation that viruses could transmit genes between bacterial cells. The second was even more unexpected: in some cases, the DNA present within viruses got incorporated into bacterial genomes. Wacław Szybalski at the University of Wisconsin-Madison Medical School, who was the first to show that a genetic defect in a cell could be repaired by transferring functional DNA from another source, introduced the phrase "gene therapy" in the early 1960s. By 1966, Edward Tatum, who had been one of the first to report bacterial gene transfer by viruses, had begun speculating:

> it can be anticipated that viruses will be effectively used for man's benefit, in theoretical studies in somatic-cell genetics and possibly in genetic therapy We can even be somewhat optimistic on the long-range possibility of therapy by the isolation or design, synthesis, and introduction of new genes into defective cells of particular organs.

Only shortly afterward, in 1968, Joshua Lederberg, who had worked with Tatum in the 1940s to discover gene transfer by viruses, developed the same idea in more detail:

> An attempt could . . . be made to transform liver cells of male offspring of haemophilic ancestry by the introduction of carefully fractionated DNA carrying the normal alleles of the mutant haemophilia gene. The precedent for this type of intervention would be the virus-mediated transduction of genetic characteristics . . . The proposal . . . would require the discovery or artificial formation of cryptic viruses to which specified genetic information relevant to the cure of genetic disease has been grafted. These viruses would then carry that information into the requisite cells of the host.

In the wake of CRISPR-mediated gene editing in somatic cells, Lederberg's speculations seem particularly prophetic.

By 1969, Robert Sinsheimer was advocating "designed genetic change," arguing that it would soon be possible to cure diseases such as diabetes through gene manipulation using viruses for transport. He recognized that the crucial problem was that of delivery: finding a suitable virus to take a gene to its intended cell. But, he thought, this was a technological barrier that would soon be crossed. Using designed genetic change to eliminate diseases was easy to defend. But, Sinsheimer also warned: "The larger and the deeper challenges, those concerned with the defined genetic improvement of man, perhaps fortunately are not yet in our grasp; but they too are etched clear upon the horizon. We should begin to prepare now for their reality."

Indeed, by 1970, the possibility of genetically altering humans had permeated popular culture to such an extent that socially responsible geneticists such as Jon Beckwith began calling for regulation. In response, in 1970, Bernard Davis published a spirited defense of unfettered genetic research, including work on human cloning. This piece was also prophetic and reads like one that could have been written at the dawn of the CRISPR era. Though Davis did not use the word "eugenics" even once, he explicitly envisioned the intentional alteration of the human germline. Recognizing the difficulty of "the incorporation of externally supplied genes into human cells," he observed that it may be easier to deliver designed genes to germline cells in an embryo than to somatic cells of specific diseases organs later on in life. As we shall see, Davis was correct.

Davis envisioned a mechanism startlingly similar to CRISPR. Discussing the use of mutagens to edit genes directly, rather than replace them (as in conventional gene therapy), he observed that "it would probably have to be attached first to a molecule that could selectively recognize a particular stretch of DNA." He was not even the first to suggest this possibility. Yet, another stalwart of early molecular biology, Salvador Luria, had broached the idea of such a mechanism earlier but had not developed it in any detail; in contrast, Davis embedded it in a profound discussion of gene editing and therapy. Most importantly, he realized that, while many diseases were affected very strongly by single genes, complex human behavioral and physical traits that could be targeted for enhancement were affected by a large number of genes and also by a wide variety of environmental factors. He dismissed what we now called genetic enhancement, arguing that they would "remain indefinitely in the realm of science fiction." Thus, by 1970, gene editing and therapy had been conceptualized quite clearly. The technology, though, lagged far behind.

FOLLOWED BY STUNNING FAILURE

Gene therapy had already been attempted by 1970; it used a virus and it failed. During the late 1950s and 1960s, Stanfield Rogers and several collaborators had been studying the Shope papilloma virus at Oak Ridge National Laboratory in Tennessee. This virus caused warts on rabbit skin when it came into contact with it but otherwise apparently had no harmful effect even when fed to the animals. Rogers and his collaborators reported a strange observation about this virus: besides the warts, it seemed to produce an enzyme called arginase in the skin cells of rabbits and that this arginase was different from that found in rabbit livers. (Arginase catalyzes the digestion of the amino acid arginine in the urea cycle, an important physiological reaction in mammals that enables the excretion of unwanted nitrogen from the body through the production of urine.) Rogers became convinced that the genome of the Shope papilloma virus contained a gene that specified arginase and that this gene was transferred into the genomes of the rabbit skin cells. A variety of observations seemed to support this inference.

Sometime later, Rogers came to hear of three sisters from a small West German village, all patients of a Köln pediatrician named H. G. Terheggen. All three sisters suffered from a very rare disease called hyperargininemia which consisted of a defect in the urea cycle of the liver that resulted in an inability to digest arginine. The result was devastating for those who suffered from this defect: it included mental retardation, developmental delays, and seizures. There was no known treatment that did any good for the patients. In 1970, Rogers and Terheggen decided to collaborate and, in the first clinical study ever of gene therapy, treat two of the girls by using the Shope papilloma virus to insert an arginine-producing gene (supposedly present in the virus) into the girls' somatic cell genomes. The two girls were duly injected with large quantities of the virus. The third, who was then an infant, was also injected some time later. The girls did not get any better. The treatment had no effect on the course of the disease; by 1975, the team had to admit total failure.

In retrospect, there is much that is ethically troubling about this episode even though the team had not broken any law or regulation in force at the time: guidelines for human gene therapy simply did not exist at the time either in the United States or West Germany (or, for that matter, anywhere else). The study failed for multiple reasons, not the least of which is the Shope papilloma virus does not contain a gene for an arginase. What is troubling about the actions of the research team was that the evidence of the presence of this gene in the virus was not conclusive in 1970 and that should have been reason enough not to attempt such a human trial. However, the clinical argument in favor of the attempt was that there was no alternative treatment at all for a very aggressive disease. The clinical argument triumphed over the procedural concerns.

RECOMBINANT DNA MADE NO DIFFERENCE

The early 1970s witnessed a dizzying sequence of technical advances in molecular biology that all made the project of therapeutic gene editing increasingly more compelling. These advances resulted in a set of techniques, collectively known as recombinant DNA. These techniques enabled the systematic editing of DNA strings and their assembly into composites irrespective of the origin of each string. Thus, artificial DNA strings consisting of genes from multiple species could be made and inserted into viruses, bacteria, and other cells. The power of these techniques worried biologists, so much so that in 1975 in the United States, a large number of the country's most prominent molecular biologists organized a conference in Asilomar, California, to discuss safety issues raised by the new technology. The conference also included lawyers, ethicists, and others who were supposed to help mediate the space between science and public policy.

The result was a self-imposed voluntary set of regulations including a moratorium on some experiments such as the cloning of pathological genes. Asilomar eventually became a potent symbol of the self-regulation of science. In the 1970s, the guidelines were largely successful in part because the number of laboratories capable of using these techniques remained limited to the United States and a handful of other countries. Moreover, research on recombinant DNA was largely controlled by a small club of prominent molecular biologists whose diktats were accepted as commands by the vast cadre of bench scientists who actually carried out the work. In the CRISPR era, Asilomar is often invoked in calls for self-regulation; the trouble is that CRISPR technology today is much simpler than the recombinant DNA techniques were in the 1970s and they are widely diffused throughout most of the world. No club any longer rules molecular biology and diktats issued by Northern scientists have little force in laboratories such as those in China.

A little after these restrictions were lifted the next year in 1976, Martin Cline from the University of California at Los Angeles (UCLA) became the first to attempt human gene therapy using recombinant DNA techniques in a living person. On the basis of studies in mice that only had very limited success, he attempted to carry out gene therapy on two patients suffering from β-thalassemia, a life-threatening anemia caused by a mutation in one of the alleles coding for part of the hemoglobin molecule. Bone marrow cells were extracted from two patients, one in Israel and the other in Italy, treated using a viral vector to have their genomes modified, and injected back into the patients.

Nothing of therapeutic value was achieved; the experiment made no difference to either patient's condition. Meanwhile, Cline and his experiment became mired in controversy. Though Cline had obtained what he thought were the necessary permissions from the relevant institutions in Israel and Italy, he had changed his experimental protocol afterward. Worse, he had never received permission from UCLA's institutional review board. Kline was censured by both the university and the National Institutes of Health (NIH) and faded into relative obscurity.

The first officially sanctioned gene therapy experiments took place in the 1990s. The two patients were children suffering from a mutation in the gene for adenosine deaminase (ADA) resulting in severe combined immunodeficiency (SCID)—the disease is known as ADA-SCID and those with it lack virtually all immune protection from pathogens. They are prone to repeated and persistent infections that can be life threatening besides suffering from pneumonia, chronic diarrhea, and extensive skin rashes. Affected children have their normal development retarded. There is no treatment making the diseases an ideal candidate for gene therapy.

W. French Anderson at NIH and several collaborators attempted gene therapy by drawing white blood cells from the patients, treating them outside the body (using a viral vector) to enable them to express the normal ADA gene, and then reintroducing them inside the body. The results were murky: while at least one of the patients showed some improvement, it was unclear whether that was a result of the gene therapy treatment or other therapies she was simultaneously undergoing.

There were many attempts at gene therapy following that pioneering attempt and, by 2020, over two thousand trials had been completed. The list of targeted diseases includes those caused by single recessive genes such as cystic fibrosis, Duchenne muscular dystrophy, and hemophilia. Because standard gene therapy adds a functional gene to the cell's genome, rather than modify the defective gene itself, it could not target dominant disease-causing alleles such as those for Huntington's disease or myotonic dystrophy. Dominance means that the presence of a single functional allele would make no difference. Gene therapy also encountered some spectacular setbacks that tarnished its image.

In 1999, eighteen-year-old Jesse Gelsinger participated in a gene therapy trial at the University of Pennsylvania in Philadelphia. He was suffering from a partial deficiency of ornithine transcarbamylase, a liver enzyme required for the removal of excess nitrogen. The functional gene was administered into his body using a high dose of adenovirus. However, the virus provoked a strong immune response and he died four days later of multiple organ failure. There was worse to come. The next year five patients in a gene therapy trial developed leukemia in response to the viral vector used.

Nevertheless, researchers have persisted in developing sophisticated methods of gene therapy hoping to cure cancer in a majority of the trials. The first gene therapy-based product to be approved was Gendicine™ in China in 2003 for treating head and neck squamous cell carcinoma, the most common form of skin cancer. However, Gendicine was approved before there had been trials that satisfied international standards and it has never been approved by the Food and Drug Administration (FDA) in the United States. Nevertheless, by 2020, there were at least five FDA-approved gene therapy products that involved gene insertion and another four that involved gene interference besides a few that are approved by the European Union but not so far in the United States.

GENE EDITING BC (BEFORE CRISPR)

By the 1990s, partly because of the slow progress of gene therapy but also because of technical advances within molecular biology, attention began to shift to devising procedures for modifying the dysfunctional genes themselves rather than bypassing them by introducing functional versions. These procedures were initially called gene "targeting" but eventually came to be known as gene "editing." These efforts took advantage of the realization that naked DNA could be introduced into many types of cells by either injecting it into the cell or simply by bathing the cell in a fluid containing that DNA and, sometimes, calcium phosphate. Somewhat surprisingly, this foreign DNA was subsequently routinely incorporated into the cell's genome.

Three developments were crucial in moving gene editing forward. Mario Capecchi at the University of Utah noticed that introduced pieces of DNA were not randomly integrated into a host cell's genome. Rather, they seemed to use homologous recombination, that is, they were integrated close to matching sequences in the genome of the host cell. As early as 1982, Cappechi and his collaborators speculated whether this observation could provide a method for targeting a gene for editing. Three years later, Oliver Smithies at the University of Wisconsin at Madison, along with several collaborators from other institutions, confirmed this speculation. In human cells derived from bladder tumors, they replaced the host's beta-globulin gene (that codes for a protein involved in cellular transport) with a version created by recombinant DNA techniques. Martin Evans at the University of Cambridge then showed that, if genes were targeted in mouse embryonic stem cells, and then these modified cells were injected back into the embryos, it was possible to produce live mice with artificially designed genes.

The trouble was that gene targeting (or "editing") was a "hit-or-miss" inefficient process with only a tiny fraction of the intended cells acquiring the modified gene. If therapy is the goal, a much more specific method had to be devised. A crucial insight was that homologous recombination, which occurred during standard chromosome duplication during the formation of germinal cells (for instance, eggs and sperm), took advantage of a double-strand break in the DNA double helix, that is, both helices were broken at the same position. A seminal model for recombination, devised in 1983 by Jack Szostak of Harvard Medical School and several collaborators including Franklin Stahl explained how: the cell would try to repair a double-strand break by trying to fuse with the homologous chromosome at the point of breakage. In ordinary cell division producing germinal cells, this led to homologous (or paired) chromosomes exchanging parts. However, if an artificial DNA segment with the right sequence was present, it would fuse with that instead as the chromosome got repaired. Thus, if a cell was supplied with an edited gene and the chromosome induced to break both DNA strands at the gene targeted for editing, the edited gene would replace the targeted one.

It took a while before someone got this mechanism to work. The first to do so was Maria Jasin at the Sloan Kettering Memorial Cancer Center in New York City in 1994. Her main innovation was to introduce into the targeted cell an enzyme that snipped DNA, forming a double-strand break, while simultaneously introducing a synthetic DNA repair template with a sequence that matched the one that was snipped. The cell's repair machinery then tppk over and incorporated the synthetic DNA sequence, including any genes attached to it, into the cell's genome.

For the enzyme, Jasin and her collaborators selected the most specific nuclease (an enzyme that cut a nucleic acid, either DNA or RNA) that they could find for the DNA target to be snipped: I-*Sce*I, the enzyme they finally used, required a precise match of eighteen consecutive DNA bases when it cut DNA. They showed that the presence of the enzyme was of critical importance. With it, almost 10 percent of mouse cells used in the experiment managed to repair the targeted DNA sequence, which was hundreds of times more than what anyone else had ever achieved before. The high specificity required by the enzyme, that is, matches at eighteen consecutive bases, ensured that only the targeted DNA was cut; that no off-target mutations were introduced. The trouble was that, to use this methodology for therapeutic purposes, the eighteen-base sequence had to be present in the targeted gene which was hardly likely. (In fact, the mouse genome did not have this sequence at all and Jasin had to introduce it artificially to perform the experiment!) The problem was to get around this debilitating limitation.

By the late 1990s, it thus became clear that a successful gene-editing technology would have to satisfy three criteria. The first is pretty obvious: it would have to include a nuclease that is capable of cutting both strands of DNA in a double-strand break to induce homology-based repair. The second criterion was that the system would accurately target that DNA sequence to cut. The third, and this was the critical problem to be solved: the system would have to be such that it could be modified at will to target any specified sequence, that is, it would have to be "programmable." In the words of Jennifer Doudna and Sam Steinberg who recounted this story from a post-CRISPR perspective:

> These next-generation gene-editing systems had three critical requirements: They had to require a specific, desired DNA sequence; they had to be able to cut that sequence; and they had to be easily reprogrammable to target and cut different DNA sequences. The first two criteria were necessary for generating a double-strand break, and the third was necessary for the tool to be broadly useful. I-*Sce*I excelled in the first two but failed miserably at the third.

It was natural to try to modify I-*Sce*I toward this end but these efforts went nowhere. Attention shifted to search for such a nuclease in nature.

Not surprisingly, searching for such a nuclease in nature proved to be futile though it beguiled several laboratories for years. There is no reason why any organism would have evolved such a nuclease. The needed breakthrough came when Srinivasan Chandrasegaran at Johns Hopkins University realized that an appropriate nuclease could be assembled in the laboratory using bits and pieces of naturally occurring protein molecules. He decided to put together part of a bacterial nuclease called FpkI that could cut DNA along with pieces from a well-known family of zinc finger proteins that recognized specific DNA sequences. Zinc finger proteins have that name because they recognize DNA using finger-like extensions arranged side by side and held together by zinc ions. These extensions consisted of repeated segments that recognized a specific DNA triplet potentially coding for an amino acid residue in a protein. Different DNA sequences could be recognized by creating these repeated segments in different ways. Once the zinc finger component of Chandrasegaran's combination molecule recognized the appropriate sequence, the Fpk1 nuclease would step up and cut the DNA. The combination molecule came to be called a zinc finger nuclease (ZFN).

Chandrasegaran collaborated with Dana Carroll of the University of Utah to develop practical uses of ZFNs for gene editing. They showed that ZFNs could be used to edit genes in frog eggs and fruit flies. In 2003, Matthew Porteus and David Baltimore used them to edit genes in human cells. Soon, the method was even used to correct a genetic defect in a human *X* chromosome. In principle, combating any human genetic disease through gene editing had become possible. Nevertheless, ZFNs never caught on: they were too difficult to program. To recognize arbitrary sequences, the zinc finger extensions of ZFNs had to be carefully tailored to combine exactly those segments that recognized the target. Only laboratories with a lot of expertise in protein engineering could do this easily. Ordinary laboratories could not.

Though ZFNs had convinced biologists that programmable nucleases were the wave of the future for gene editing, their use was short-lived.

In 2009, ZFNs were surpassed in ease of use by a new technology based on very similar principles. This technology combined a new type of protein called transcription activator-like effectors (TALEs) with a nuclease (to form a TALEN). TALEs had been fortuitously discovered in Xanthomonas bacteria that cause spots and blights in plants and were well studied because of the economic harm they caused. Like zinc finger proteins, TALEs also consist of multiple repeated segments that, together, recognize DNA but with one crucial difference: whereas zinc finger protein segments recognize a triplet of DNA nucleotides, TALE segments recognize individual nucleotides. This means that programming a TALEN to recognize any given DNA sequence is much simpler than was the case for ZFNs. In principle, only four TALE segments would suffice to recognize each of the four DNA nucleotides and they could be combined to target any DNA sequence whatsoever.

There was good reason to believe that TALENs would drive gene editing in the future. Unfortunately for them, TALENs were stillborn. By the time TALEN technology was being perfected for everyday use, CRISPR came along in 2012 using RNA rather than proteins to recognize targeted DNA sequenced for editing. As Carroll noted ruefully in 2015:

> But pity the poor TALENs. Only 3 years after the elucidation of the TALE recognition code, the CRISPR-Cas platform arrived on the scene. The remarkable simplicity of CRISPR-Cas rapidly made it a favorite in laboratories around the world. No protein engineering was required, a teenager could design new [RNAs] for new targets and multiple targets could be attacked simultaneously. The high success rate helped establish CRISPRs as the programmable nuclease of choice for research. As many people have said, this platform democratized genome engineering.

But, before, we turn to the CRISPR story, let us take an excursion into the hopes and achievements of the HGP because that is what has made CRISPR possible. From a medical perspective, the HGP has so far contributed virtually nothing in spite of all the hype surrounding it. But, by enabling safe, accurate, fast, and cheap genome editing, CRISPR may have rescued the reputation of the HGP.

Chapter 3

What Good Was the Human Genome Project?

"I think there will be change in our philosophical understanding of ourselves. Three billion bases can be put on a single compact disc (CD), and one will be able to pull a CD out of one's pocket and say 'Here's a human being; it's me!'."

—Walter Gilbert, 1992, "A vision of the Grail."

"This book tells the story of one of mankind's greatest odysseys. It is a quest that is leading to a new understanding of what it means to be a human being, and is now being carried out under the auspices of the Human Genome Project."

—Walter Bodmer and Robin McKie, 1994. *The Book of Man.*

THE SEQUENCE REVEALED

On June 26, 2000, the official HGP of the United States, together with a private venture started by Craig Venter, announced the completion of what they called a draft human genome sequence. Most but not all of the human genome had been sequenced—the gaps were glossed over in the announcement that came from the East Room of the White House. US President Bill Clinton was lyrical: "Nearly two centuries ago Thomas Jefferson and a trusted aide spread out a magnificent map [of the American West, just charted by Lewis and Clark] . . . that defined the contours and forever expanded the frontiers of our continent and our imagination. Today the world is joining us here . . . to behold the map of even greater significance. We are here to celebrate the completion of the first survey of the entire human genome. Without a doubt, this is the most important, most wondrous map ever produced by human kind."

Over satellite, British Prime Minister Tony Blair chimed in: "Ever since Francis Crick and Jim Watson . . . made their historic discovery in the middle of the last century [of the DNA double helix], we've learned that DNA was the code to life on Earth. And yet I guess for Crick and Watson, the process of identifying the billions of units of DNA and piecing them together to form a working blueprint of the human race must have seemed almost a superhuman task beyond the reach of their generation. And yet today, it is all but complete. Nothing better demonstrates the way technology and science are driving us, fast-forwarding us all into the future."

Neither Clinton nor Blair mentioned that the HGP had been intensely controversial among biologists when it was first proposed in the 1980s. Perhaps, they didn't even know. Worries about the HGP, about genomics since then, and about ubiquitous DNA sequencing as if it is an end-in-itself, have all been drowned by the hype surrounding genomics since the turn of the century. While acknowledging its contributions and importance, we will take a skeptical look at the HGP in this chapter. Its title is a paraphrase (in past tense) of a talk I used to give in 1992. A transcript was eventually translated into Spanish by the eminent Mexican biologist, Arturo Lazcano, and published in México in 1992. Otherwise, it is forgotten. Much of the remainder of this book will be concerned with whether CRISPR can rescue some of the unkept promises of the HGP.

Clinton's and Blair's enthusiasm would suggest that Gilbert's vision (in the quotation at the beginning of this chapter) was about to be fulfilled. Over US$ 3 billion had been officially spent on the project (and some estimates make the sum much higher). The official HGP, which was formally initiated in 1990, was only supposed to be completed in 2005. It was supposed to consist of a collaboration of several large laboratories, mainly in the United States and Britain, and it was supposed first to map all the human genes and then move on to sequence the complete genome. The HGP was conceived of as a massive "Big Science" project with major laboratories assigned specific chromosomes to sequence. There was supposed to be open and full sharing of all data and technological advances. Along the way, the HGP would develop new mapping and sequencing technologies to drive biology into the future. Bodmer and McKie predicted it would "transform medical practice in the next century." This sentiment was shared—at least in public—by every one of the HGP's proponents.

However, the unexpected emergence of a competing private project started by Venter in 1998 had derailed the official HGP's stately progress. Venter, who had founded The Institute for Genome Research with industry funding in 1992, announced a program to sequence the genome in three years at a fraction of the cost of the official HGP. The private funding interests were motivated by the hope of creating and exploiting a massive database linking genomic variation with medically relevant features. Venter and his commercial backers expected to privatize all data and patent every result when possible, and then reap the monetary benefits.

In response, the official HGP rushed work to complete its own version of a draft sequence. In 2000, it claimed to have finished this task and announced a final product simultaneously with Venter. Skeptics would point out that the final "draft" had been put together rather hastily with almost 10 percent of the sequence still to be completed. Some of these unfinished fragments were sequenced by 2006; still, as of 2010, a "whole" human genome for any individual constituted only about 93 percent of the full sequence—the other 7 percent had remained too hard to sequence. Even today, though the percentage of the sequenced genome has increased, the entire human genome has not been completed; in 2018, 875 major gaps still remained to be completed. The situation was only marginally better in 2020. We will encounter many other similar problems later, including what it means to talk of the sequence of *the* human genome.

However, these gaps have never prevented enthusiasts of the HGP from extolling its scientific and medical virtues. One year before the announcement of the draft sequence, Francis Collins, then director of the (United States) National Human Genome Institute promised an "individualized medicine" (more commonly known as "personalized medicine") and painted the following rose-tinted scenario for the future of medicine: "John, a twenty-three-year-old college graduate [in 2010], is referred to his physician because a serum cholesterol level of 255 mg per deciliter [which is higher than normal] is detected in the course of a medical examination required for employment." Using an interactive program that takes into account John's habits and family medical history, the physician suggests a battery of genetic tests. John agrees to fifteen tests for diseases that have preventive interventions and rejects ten others that do not. These tests rule out many diseases but John is then

> sobered by the evidence of his increased risks of contracting coronary artery disease, colon cancer, and lung cancer. Confronted with the reality of his own genetic data, he arrives at that crucial "teachable moment" when a lifelong change in health-related behavior, focused on reducing specific risks, is possible. And there is much to offer. By 2010, the field of pharmacogenomics has blossomed, and a prophylactic drug regimen based on the knowledge of John's personal genetic data can be precisely prescribed to reduce his cholesterol level and the risk of coronary artery disease to normal levels. His risk of colon cancer

> can be addressed by beginning a program of annual colonoscopy at the age of 45, which in his situation is a very cost-effective way to avoid colon cancer. His substantial risk of contracting lung cancer provides the key motivation for him to join a support group of persons at genetically high risk for serious complications of smoking, and he successfully kicks the habit.

There will be more to say about John later in this chapter. What matters most in our context is that Collins' hope was that the budding new field of pharmacogenomics would use John's sequence data to tailor his drugs to his individual sequence.

As Collins put it: "Identifying human genetic variations will eventually allow clinicians to subclassify diseases and adapt therapies to the individual patient. There may be large differences in the effectiveness of medicines from one person to the next. Toxic reactions can also occur and in many instances are likely to be a consequence of genetically encoded host factors." By 2015, this desire would be the basis for Collins' promise for precision medicine. But, it would still be a project for the future. It is 2020 now. To the best of my knowledge, no physician anywhere in the world has ever requested a patient's complete DNA sequence to devise a medical intervention. Personalized medicine still remains a hypothetical project for the distant future with little reason to believe that it will ever happen. In the rhetoric of those who promote a genome-based medicine, it has been replaced by a somewhat less ambitious promise of a "precision" medicine without explicitly acknowledging that the dream of a personalized medicine by 2010—or even by 2020—has proved to be an illusion. Indeed, little has changed since 2000.

COMMON DISEASES AND COMMON VARIANTS

In the 1990s, Collins' hopes were perhaps not entirely unrealistic. What the optimists were relying on was a plausible hypothesis, the common disease-common variant (CD-CV) hypothesis that claims that common diseases were controlled by statistically detectable common variants of genes (where, typically "common" is taken to mean that an allele occurs in a population with a frequency higher than 5 percent). Every human being was supposed to have some of these common variants and, thus, intervention strategies could potentially use them to devise different treatments tailored to each individual.

Most of these common variants were supposed to have been generated very early in human evolution. They were supposed to have become fixed (that is, present in all members) in the small human populations that then existed before these groups went through explosive population expansion. If many such genes are involved in causing some disease, but each by itself has a small effect, there is an evolutionary reason why they would persist in modern human populations. Because each such gene has a small effect, a person with such an allele would only have a very slightly lower reproductive fitness compared to someone without that allele. This means that these individuals would leave only a few less offspring relative to others in the group. Thus, natural selection would not be able to weed this allele out of the population effectively compared to what it could do to a disease-causing gene with a large effect and, therefore, making a big difference in fitness. We would each be carrying some of these disease-implicated common variants. By the late 1990s, by spawning technologies for fast whole-genome sequencing, the HGP had made it possible both to test the CD-CV hypothesis and potentially use it (if correct) for therapeutic purposes.

Indeed, the development of these fast sequencing techniques for large genomes, a very important contribution of the HGP, enabled the collection of large data sets of genome sequences from a large number of individuals. These data sets made it possible to study statistical associations between allelic variants and common diseases. What we look for is simply how often one of these common allelic variants co-occurs in a person along with any trait that we are studying. A major use of these genome-wide association studies (GWAS), as they are called, was to study variation at the level of individual nucleotides distributed throughout the genome. These are called single-nucleotide polymorphisms (SNPs) because they show variation (that is, polymorphism) at the level of individual nucleotides. The statistical association between these SNPs and diseases could then be used to identify even large numbers of genetic variants that may potentially play a role in causing the disease.

In 1998, a newly minted SNP Consortium began assembling maps of SNPs across human genomes. It was followed by the HapMap project which started in 2002 and began mapping sets of SNPs found near each other (and called a "haplotype"). By 2010, hundreds of common SNPs had been associated with several diseases including Alzheimer's disease, hypertension, schizophrenia, and type 2 diabetes. However, do these associations mean anything? Skeptics pointed out that these associations were proving to be useless when it came to predicting disease. For instance, in the cases of type 2 diabetes, statistical association studies of 2.2 million SNPs in more than ten thousand people have identified 18 SNPs associated with the disease; yet, taken together, differences between nucleotides at these sites explain at best only six percent of the variability in the onset of this disease.

The absence of successful predictions suggested that the CD-CV hypothesis may not live up to what it had been trumped up to be. In fact, the CD-CV hypothesis had always been controversial ever since it was first proposed in the 1980s. An alternative common disease-rare variant (CD-RV) hypothesis claimed that common diseases were the result of rare genetic (allelic) variants with large effects (to the extent that genes were involved at all). Many individually rare variants would be implicated for the same disease (which is why the disease is common). Association with these rare variants cannot be detected by the statistical analyses of GWAS. The statistical tests would not have what statisticians call sufficient power: the variants are so rare that their co-occurrence with the traits would not be deemed statistically significant. Because they only measured statistical associations, GWAS would churn up a large set of common variants which were, indeed, common and present—but had no medical relevance. There is plenty of evidence for the CD-RV hypothesis including genes for predisposition to breast cancer, genes for disease of lipid metabolism, and some for severe mental illnesses. Paraphrasing Tolstoy in *Anna Karenina*, the biologist, Kenneth M. Weiss, of Pennsylvania State University observed: "All healthy families resemble each other; each unhealthy family is unhealthy in its own way."

Those in favor of the CD-RV hypothesis (over CD-CV) include prominent British human geneticist, Walter Bodmer, who was quoted at the beginning of this chapter. If CD-RV is correct, all that GWAS have revealed is a great deal of DNA heterogeneity that is causally irrelevant to disease. That explains why good predictions do not emerge from these results. This means that, to the extent that the medical goals of the HGP were to be reached through these studies (that is, through GWAS), it should come as no surprise that, as of 2020, those goals have not been achieved. What, then, has the HGP contributed to medicine? As far as generating new treatments was a goal, it is hard to escape the conclusion that, as of 2020, it has contributed very little.

MEDICAL IRRELEVANCE OF THE SEQUENCE

Let us return to 2010 when (in Collins' story from 1999) John is supposed to have had his physical examination. In reality, John would have encountered very little in the physician's office that was different from 2000. There was as yet no country in the world with routine preventive genetic testing for twenty-three year olds. There was no expanding field of pharmacogenomics. Writing in *Scientific American* in 2010 about the medical contributions of the HGP, Stephen S. Hall observed:

> the scientific community finds itself sobered and divided . . . The problem is that research springing from the genome project has failed as yet to deliver on the medical promises that Collins and others made a decade ago. Tumor biologist Robert A. Weinberg of the Whitehead Institute for Biomedical Research in Cambridge, Mass., says the returns on cancer genomics "have been relatively

> modest—*very* modest compared to the resources invested." Harold E. Varmus, former director of the National Institutes of Health, wrote recently in the *New England Journal of Medicine* that "only a handful of major changes . . . have entered routine medical practice"—most of them, he added, the result of "discoveries that preceded the unveiling of the human genome." Says David B. Goldstein, director of the Center for Human Genome Variation at Duke University: "It's fair to say that we're not going to be personalizing the treatment of common diseases next year."

Even a decade later, in 2020 there is no reason to change this assessment. And it is a good bet that there has been no person who has walked up and presented a CD containing a genome sequence to a physician and announced, "It's me."

Of course, smokers like John are advised to quit and join support groups, if helpful and available, to decrease the risk of getting lung cancer. But that has been going on since the 1960s with incrementally increasing knowledge of the mechanisms by which tobacco smoke induces lung cancer. It did not require a completed human genome sequence. Tobacco companies have long sought evidence for a genetic predisposition to nicotine-induced cancer. They have done so with the hope of avoiding liability arising from having intentionally sold an addictive carcinogen, sometimes intentionally doctored in such a way as to make the product more carcinogenic. In spite of the human genome sequence, and despite lavish funding of this research, any evidence of genes associated with smoking-related cancer has remained illusory. But this is what Collins seems to have expected when he painted the scenario of John's diagnosis and treatment. It was wishful thinking.

In most Northern countries, newborns are screened for a wide variety of disorders. In the United States, the list of these disorders varies by state. In Texas, newborns are screened for serious heart defects, hearing, and fifty-three other disorders many of which are genetic including cystic fibrosis, phenylketonuria, sickle cell trait, and thalassemia. Not one of these tests has its origin in the results of the HGP. Rather, the list has incrementally increased over decades using results of traditional medical research.

CRITIQUES OF THE HGP, PAST AND PRESENT

This apparent medical irrelevance of the results of the HGP plays into the hands of the skeptics who questioned the HGP when it was first proposed in the 1980s. It will be worth our while to revisit those arguments because some of them, particularly those questioning the hegemony of the gene, are also relevant to today's arguments about CRISPR. As noted earlier in this chapter, in the 1980s, the proposal to sequence the entire human genome generated immense controversy both within and outside the biological community.

Critics came in various stripes. There were critics who thought it was wrong to sequence the human genome at all. Then, there were those who thought that massive blind sequencing was bad science policy. The HGP consisted of *blind* sequencing because it consisted of systematic mechanical sequencing the entire genome without any attention to function. In particular, sequencing was not to be restricted or even primarily targeted to known or presumed genes. No one questioned the value of sequencing genomic regions containing genes. Since about 95 percent of human DNA was not known in the 1990s to have any functional role at all, much of such blind sequencing was going to consist of snipping away at functionally irrelevant or "junk" DNA. There will be more on junk DNA below. There were also a large number of critics who were wary of past eugenic abuses and worried about the potential misuse of sequence data. Finally, there were those who doubted the scientific and medical value of the HGP—their arguments are the ones most relevant to us because they carry over to our context.

Turning briefly to those who thought that it was wrong to sequence the human genome, for many of them, humans had no business with such knowledge. Most such critics brought religious beliefs along with them: HGP proponents were apparently embarking on a path to forbidden knowledge; they were playing "god." As Robert Cook-Deegan put it in 1991:

> "Nothing arouses public discomfort with the future of genetics as much as visions of scientists in white coats mucking about with the genes of future children. A deep distrust of elites and errant technological prowess lurks just below the surface, evoking images of Frankenstein and the Golem . . . and a spate of science fiction misadventures project dark visions of a future governed by scientists and technicians bereft of emotional depth and moral judgment."

Even earlier, in the context of the original invention of recombinant DNA techniques, in the United States, the President's Commission for the Study of Ethical Problems in Medicine and Biomedical and Behavioral Research dedicated a section of its report, *Splicing Life* of 1982, to discussions of "playing God." It worried the Commission.

In a similar vein, also in 1992, the European Parliament asserted: "the rights to life and human dignity . . . imply the right to inherit a genetic pattern that has not been artificially changed." However, the right was qualified so as to "not impede development of the therapeutic applications of genetic engineering (gene therapy), which holds great promise." Cook-Deegan went on to note:

> Dozens of petitions to proscribe germ line gene therapy have circulated, most arguing from one of two streams of thought. One set of arguments rests on natural law which is based on theological traditions that place humans in a special

> category. Human DNA is elevated to sit on that same pedestal. Another set of arguments trace to the Greek notion of hubris. We cannot be trusted to know what we are doing with such powerful technology.

Cook-Deegan was unsympathetic to such human exceptionalism and left open the possibility of conscious genetic alteration of populations. But we should remain aware of these arguments because they remain fully relevant to our post-CRISPR world.

The point that deserves emphasis is that the technological innovation of new gene-editing methods by itself does not address whatever merit these old arguments had. Using CRISPR-based techniques to edit the human germline is precisely the type of action that these critics have always viewed as "playing God." Even if we have no desire to become entangled in religious disagreements, if we are willing at all to countenance human germline editing, it would be wise policy to engage these arguments as rationally as possible and assuage worries about "playing God." We will try to do so in the sixth chapter.

There were many who thought massive blind sequencing was unwise science policy. These critics accepted that, sooner or later, the human genome would be sequenced and, unlike their more religious counterparts, they were not concerned about acquiring this knowledge. While a majority of these critics probably thought that the results would be scientifically and medically useful, not all of them were convinced of their value, as we shall see below. But even the optimists had worries. Three of these worries were persistent. The first of these was the cost of the project: complete genome sequencing was going to be expensive. Critics wondered about how this would affect funding for the rest of biology. At stake was the massive funding that would be dedicated to sequencing alone during the planned fifteen-year tenure for the project. Would the project siphon money away from basic science to what critics perceived to be a purely technological goal of large-scale blind sequencing?

A second worry was there that the project, as an emblem of Big Science, would change the ethos of biological research. Biology would no longer be based on individual laboratories centered around each scientist. Critics claimed that such a development would harm the ethos of biological research. However, the critics offered little compelling reason as to why collaboration and teamwork, on which the HGP would rely, was going to be so harmful for the future of biology. Finally, the HGP also differed from twentieth-century biology in yet another way. Much of biology during that century had embraced hypothesis-driven research: every experiment was supposed to test a hypothesis. There was a stark contrast with the HGP: blind sequencing does not refer to any hypothesis at all. With the HGP, critics feared, biology would change beyond recognition—and not change for the better.

But, once again, there was good reason to be skeptical about the critics' concerns. Yes, twentieth-century biology had indeed become hypothesis-driven but there is more to science than the testing of hypotheses. There is the wonder and joy of exploration, for instance, what the great European naturalists Alexander von Humboldt and Alfred Russel Wallace experienced when they wandered around tropical rainforests at two ends of the world, Humboldt in the Amazon and Wallace in the Malay archipelago, cataloging the plants and animals they found. The HGP was similar except that it was inward bound, proposing to explore the genomic terrain within rather than the forests and rivers around us.

There were many critics who were worried about how sequence data would be used but did not believe that the human genome should not be sequenced at all. The HGP acknowledged the seriousness of this problem as we saw in the last chapter. From the outset, through the Ethical, Legal and Social Implications (ELSI) Program, the HGP devoted 3 percent of its budget to analyze and prepare responses to social problems. The critics were worried that society was ill prepared to cope with problems that would be created by the widespread availability of human sequence data. They were most concerned with the proposed speed of the project: that the human genome should be sequenced so rapidly. The pace of the crash program to sequence the human genome was viewed with unease because there would not be enough time to prepare for the potential societal consequences of having the data.

For example, genetic counseling is supposed to inform potential parents about the likelihood of a child to inherit some genetic disease that runs in families; ideally, it is also supposed to inform expecting parents about the genetic constitution of a child about to be born and what it means. Genetic counseling is often difficult—and ineffective—because the laws of transmission of genes from parent to child are probabilistic: they only predict what is likely to be inherited and to what extent. All claims about potential genetic susceptibilities to disease are also probabilistic. These are not everyday concepts that most potential parents can readily understand. Genetic counselors try to guide them through the relevant complexities. It was clear that, as more and more information emerged about the human genome sequence, there would be increasing demands for genetic counselors. In the United States, it was believed that not enough of them were being trained in the late 1980s. There would potentially be a genetic counseling crisis in the 2000s after the full sequence became known. Now, without the HGP, it would probably have taken about a generation (roughly 30–40 years) for the full human genome sequence to become available, sometime between 2020 and 2030. That would give a fair amount of time to ensure that an adequate supply of genetic counselors would be available. But the HGP reduced the time frame to fifteen years. Would that be sufficient to satisfy the needs of genetic counseling? Was this wise policy?

Yet, others argued that the knowledge obtained would be dangerous. Employers, insurance providers, and others could discriminate against individuals on the basis of their sequences. We could end up with a class of "asymptomatically ill" individuals : people who would be regarded as ill because of sequence data that purportedly showed them to be susceptible to some disease even though they had no symptom of the disease. In the United States, we could end up creating a "biological underclass" in a society in which socialized medicine was deemed inconceivable. Individuals could also be stigmatized because of some genetic propensity. One identical twin could reveal much about the other's health without that twin's consent. It did not even matter whether the genome sequence really had such power: as long as society perceived that it did, the problems would materialize. (As many of us suspected back then, and has been amply demonstrated post-2000, the human genome has no such predictive power.)

These concerns were real enough and remain pertinent today even after considerable regulatory innovation to address at least the problem of discrimination. However, even back in the 1980s it was realized that there could be three different responses to these problems: we could choose not to sequence the human genome at all—this is the option that was also preferred by the religious critics mentioned earlier. We could sequence, slowly, without the crash project of blind sequencing envisioned by the HGP; meanwhile, we could try to prepare for the problems that were expected to emerge. Or, third, we could proceed with the HGP but also address the problems simultaneously. The last two of these options treat the question as if it were one of devising appropriate policy; only the first embraces the stronger view that we should not sequence. By pursuing the HGP in the way that it did, the United States took the third option.

In the United States in the early 1990s, long before the Affordable Care Act of 2010 assured insurance coverage for preexisting conditions, the potential for genetic discrimination by health insurers was immense: gene sequences supposedly conferring susceptibility for a disease amounted to a preexisting condition even when no symptom of the disease had manifested itself. Insurers could discriminate on those grounds; employers had done so even using earlier technologies. Two well-known cases had occurred before protective measures were finally in place. In 1998, it became publicly known that the Lawrence Livermore National Laboratory at Berkeley had secretly tested workers from the 1960s to 1993 for sickle cell disease, syphilis, and pregnancy without their knowledge or consent. The workers thought they were being screened for cholesterol. In 1999, the laboratory settled with its victims. In 2002, the Burlington Northern Santa Fe Railway company paid US$ 2.2 million to thirty-six workers to settle a case in which it was accused of secretly testing them, after they had filed work-related injury claims, for carrying a DNA variant that was supposed to predispose individuals to carpal tunnel syndrome. Those claims of genetic predisposition for carpal tunnel syndrome turned out to be incorrect but that does not mitigate the ethical and legal issues raised by this attempt at genetic discrimination.

However, in 2008, after more than a dozen years of haggling, the United States Congress finally passed the Genetic Information Nondiscrimination Act, which US President Bush signed into law. Its passage addressed many of the worries that had been raised by critics of the HGP even though it left many holes. While it outlawed sequence-based discrimination in employment and insurance, it did not address life, disability, or long-term care insurance. In 2010, to a large extent, the Affordable Care Act addressed these limitations. The Affordable Care Act has continually been under threat from Republicans and the administration of Donald Trump did much to take the teeth out of it. But its ban on insurance denial because of preexisting conditions remains popular. As long as the present legal situation remains unsullied, we have good reason to claim that the United States has been quite successful in its strategy of sequencing the human genome while simultaneously addressing the social and ethical problems that are emerging.

Of course, the United States had been somewhat unique among Northern nations in not having universal health care coverage legally mandated until 2010; most European countries moved toward universal coverage shortly after World War II and, in some regions such as parts of Germany, these policies go back to the nineteenth century. In general, most developed nations do not face problems posed by an absence of comprehensive health care. The situation remains murky in China and India, two other large countries with the technological wherewithal to use DNA sequences. Given the Chinese embrace of CRISPR, which we will discuss in some detail in later chapters, what happens there may be unique and interesting.

Finally, there were those critics who doubted both the scientific and medical value of the project. Molecular and evolutionary biologists focused on different sets of problems. Molecular biologists worried about two issues: when it comes to making predictions about an organism's structure or behavior, what good would the sequence do?; and what was the point of sequencing so much junk DNA? Gilbert was convinced that the availability of complete sequences for a variety of organisms would jumpstart a new theoretical biology. The sequences would allow us to predict how an organism would look and behave. Probably because he was yet another ex-physicist who had made seminal contributions to molecular biology, Gilbert was unduly swayed by the potential to make predictions. If, indeed, it turned out that way, biology would emerge as a predictive science on par with physics. But how likely was such an eventuality?

Gilbert's critics noted how little of the phenotype seemed to be specified by the genes. Even identical twins had different personalities and behaviors. Most importantly, in us, antibodies for a crucial part of our immune system and are specified by genes. Identical twins need not have similar immunological profiles because they have different genes for antibodies because of ways in which the genome created at fertilization diverges during the formation of the cells that produce antibodies. (Contrary to common belief, not every cell of our body has an identical genome. Many cells of our immune system do not.)

Yes, genes allow the prediction of an extra finger or a cleft lip. These are no doubt spectacular phenotypes controlled almost entirely by the action of individual genes but such phenotypes are very rare. Even if the HGP would allow a better understanding of some complex phenotypes such as hypertension or obesity, this was a far cry from a new predictive theoretical biology. Gilbert did not convince the skeptics. The explosion of work on epigenetic factors since 2000 only validates this skepticism: all kinds of molecules that attach to the DNA component of chromosomes, and respond to environmental signals, control what the DNA sequence can do. Sequences alone tell very little about an organism's phenotypic features.

The most that a DNA sequence specifies through the genetic code is the amino acid residue sequence of a protein. Parts of the DNA sequence are also used to regulate when other parts are used to produce proteins in this way—these parts are called regulatory sequences. However, except for bacteria and similar simple organisms, only part of the genomic DNA plays either of these roles; the rest, as far as we know, do not do anything useful at all. As we have noted earlier, in humans, it has been known since the 1980s that an estimated 95 percent of the DNA seemed to do nothing at all—it had been classified as "junk" as the evolutionary geneticist, Susumu Ohno put it in the early 1970s. And yet, critics pointed out, the HGP would spend billions of dollars blindly sequencing such DNA.

The contrast here is between blind and targeted sequencing—the latter referring to sequencing chunks of DNA that were known to contain genes, that is they either specified proteins or played a known role in gene regulation (turning genes on or off). Targeted gene sequencing had been around for decades and, as we also noted earlier, no competent biologist denied its importance. But, even here, there were problems with the creation of a predictive biology. Proteins were clearly the most important molecules in organisms; as we saw in the last chapter, their sizes and shapes (what molecular biologists called their three-dimensional conformations or tertiary structures) determine their biological functions, for instance, whether an enzyme is able to digest a nutrient molecule, or an immunoglobulin molecule of the immune system is able to remove a foreign substance.

But DNA sequences do not directly specify the shape of a protein; they only specify its amino acid residue sequence—the linear or primary structure of a protein. The task of predicting the shape of a protein from its amino acid residue sequence is known as the protein folding problem: it has been recognized as extraordinarily difficult since the 1960s. The protein folding problem remains unsolved even today and it now seems extremely unlikely that the amino acid sequence of a protein by itself specifies the shape of every protein. The physical process of synthesizing a protein at the ribosome may also play a role. Thus, according to these critics, the gap between a genome sequence and the biology of a cell, let alone an organism was far too wide to be likely to be bridged by 2005 when the HGP was supposed to be completed. The protein folding problem was one of many others that stood in the way.

But, perhaps, there can be a way out. The three-dimensional structure of a protein can also be experimentally determined using crystallography. If the sequences of two proteins are similar, at least in part, and the shape of one is known, we may infer part of the shape of the other by exploiting sequence similarity. In a spirited defense of his vision for HGP in 1992, Gilbert emphasized the power of this strategy. The trouble is that even today not enough protein structures have been solved to play this game with much confidence. Determining the structure of a protein using crystallography is much more time consuming than sequencing its gene. By and large, biologists interested in any structures from the level of cells to whole organisms did not see how genome sequences would provide a bonanza to their disciplines.

Evolutionary biologists were perhaps even less impressed. Jim Crow from the University of Wisconsin at Madison, a dominant figure in evolutionary theory, pointed out that it would be of greater scientific value to know 10 percent of the genome of ten species rather than the entire genome of any one species. Here, at least, the HGP had an adequate response: early on, it was decided to sequence simpler genomes of several other species to hone technological skills on the way to sequencing the human genome which was expected to prove more recalcitrant.

Evolutionary biologists also pointed out that the whole idea of "*the* human genome" was incoherent. Except for identical twins, no two human beings have genomes that are the same. (Even in the case of identical twins, as we saw earlier, the genome of some cells, for instance, some of those forming the immune system are different.) Given the amount of possible variation, it is reasonable to believe that there are as many "normal" human genomes as there have been human beings—"normal" in the sense that the individuals with these genomes would not be suffering from debilitating illnesses because of their sequences. What, then, is *the* human genome? In the HGP, there seems to have been an implicit assumption that much of the genome would be identical for almost all human beings. From an evolutionary perspective, there seems to be ample reason to believe that the HGP was fundamentally ill-conceived.

A slew of recent results underscores the scope of this problem. Over the years, since the publication of the first draft sequence in 2000, biologists have completed, revised, and refined what they call a "reference" genome. (This genome is supposed to be the yardstick against which potentially functionally abnormal sequences are compared.) Though biologists hope that this reference standard reflects human diversity, and there are many ongoing attempts to make sure that it does, 70 percent of the reference genome comes from a single individual. A 2017 paper estimated that there would be about 16 million DNA base pair differences between the reference genome and that of a random individual.

A 2018 paper by Steven Salzberg and Rachel Sherman (along with forty-four collaborators) showed how odd the reference genome is. They compared the sequences of 910 individuals, all of African descent and from twenty different countries, to the reference genome. They found 300 million DNA base pairs that were common to all 910 individuals but entirely absent from the reference genome. It is enough DNA to form an extra chromosome. In general, DNA sequences belonging to Africans are about 10 percent larger than that of the reference genome. Even more recently, a 2019 paper analyzing 154 genomes from twenty-six ethnically different populations from around the world found 60 million DNA base pairs that were missing from the reference genome. In the past few years, there have been concerted efforts to address the problem of human diversity at the genomic level through multiple projects with the goal of producing a diverse array of reference genomes rather than the single one hypothesized and subsequently canonized by the HGP. The critics seem to have been prescient in objecting to the very idea of *the* human genome.

It should be emphasized that the inability to use DNA sequences to predict functional biology at the level of the organism carries over to the medical context. There, it means that therapies cannot be devised from sequence information alone. Recall the example of sickle cell disease from the last chapter. We know the three-dimensional molecular structure, that is, the shape and size of the mutant hemoglobin in great detail. Yet, we have been unable to design effective therapies to manage the disease. What does knowing the sequence of the mutant gene change? Nothing, and this sequence has been known since the 1970s. From the DNA sequence, we can predict the mutant amino acid sequence of sickle cell hemoglobin. We cannot even predict the known three-dimensional conformation: that is the bite of not having solved the protein folding problem. In the case of sickle cell disease, having a completed human genome sequence takes us no further toward designing effective therapy than where we were since the 1950s.

EVOLUTION AND ARCHITECTURE OF THE GENOME

However, at least as far as basic science is concerned, the critics of the HGP can be answered but in a rather odd way. The critics were correct about what they predicted: the HGP did not deliver on any of its *explicit* scientific promises. But that turns out to be completely irrelevant. No new theory-driven molecular biology has emerged. The protein folding problem is no closer to solution today than it was in 2001. We are in no position to predict functional biology (how organisms function as a whole) from DNA sequences. Yet, probably no one well versed in biology can deny that the publication of the draft human genome sequence was one of the major events in the history of science. The reason? The human genome was full of surprises. Almost every assumption that biologists had previously made about the genome and its evolution turned out to be false. *That* is the striking scientific contribution of the HGP and its influence on biology cannot be over-estimated.

By 2001, when the draft sequence of the human genome was published, besides thirty-nine bacterial species, the complete genomes of baker's yeast, *Saccharomyces cerevisae*, the worm, *Caenorhabditis elegans*, and the fruit fly, *Drosophila melanogaster*, had already been sequenced. Since then, eukaryotic whole-genome sequences continue to be reported at a steady rate. The largest eukaryotic genome recorded so far seems to be that of an endemic plant from Japan, *Paris japonica*, that has 150 billion base pairs. While this genome is yet to be fully sequenced, the smallest recorded nuclear genome, that of the intracellular parasite, *Encephalitozoon intestinalis*, has been sequenced and found to contain only about 2.3 million base pairs.

In 2001, the biggest surprise from the completed human genome sequence was the extraordinarily low number of genes. In the 1990s, while Gilbert had put 300,000 as the upper limit of the possible number of human, most estimates ranged between 60,000 and 140,000, with the 1990 plan for the HGP embracing an estimate of 100,000. Instead, the completed sequence suggested about 30,000–40,000 genes. Since then, this estimate has decreased to 20,000–25,000, with more recent estimates hovering around 22,500. The same estimate holds for the mouse, *Mus musculus*, and is not much more than the 21,200 gene number estimate for the worm, *C. elegans*. The fruit fly, *D. melanogaster* has 16,000 genes which is only a little lower. Meanwhile, the mustard weed, *Arabidopsis thaliana*, has 25,000 estimated genes but rice, *Oryza sativa*, has as many as 60,200. The pufferfish, *Fugu rubripes*, has 38,000 genes. There is no straightforward relationship between gene number and the complexity in the structure or behavior of the organism.

This paradoxical lack of correlation between perceived complexity and their gene number has been called the G-value paradox by puzzled biologists. The number of genes is also not correlated with genome size. The original report on the sequence also noted that the human "proteome" or protein set is much larger (and, in that sense, more complex) than that of invertebrates. This puzzle is resolved by the higher prevalence in humans of what is called alternative splicing. Splicing is the process of removing noncoding stretches (introns) of messenger RNA transcribed from a gene. It occurs within the nucleus at an organelle called the spliceosome before the modified messenger RNA leaves the nucleus gets translated into protein sequences at ribosomes in the cytoplasm.

During alternative splicing, different stretches get removed from different messenger RNA transcripts from the same gene. This results in multiple final messenger RNA pieces that get translated into different proteins, all emerging from the same gene. According to recent estimates, more than half of the human genes are subject to alternative splicing with an average of 2.6 transcript variants per gene; in contrast, only 20 percent of the genes are alternatively spliced in *C. elegans* and *D. melanogaster*, with an average of 1.3 transcript variants per gene.

There were other surprises in the complete human sequence of 2001. The original report claimed that there had been horizontal gene transfer of hundreds of bacterial genes into the human genome; however, this high estimate did not survive further analysis with more recent estimates putting the number at around forty. The distribution of human genes between the chromosomes and within them was highly uneven. Human genes tend to occur in clusters. The human genome has about four thousand pairs of duplicate genes and 5 percent consists of recently duplicated segments. Almost a third of the genes in the human genome appear to be "orphans," that is, they have no homolog (a similar gene inherited from a common ancestor) in any other non-primate species. The human genome also has about fifteen thousand pseudogenes which are imperfect nonfunctional versions of working genes. In 2001, only about 2 percent of the human genome was estimated to specify amino acid sequences; since then that estimate has come down to 1 percent. Within each gene, on the average, there is thirty times as much junk DNA as functional DNA. While reliable estimation of the amount of regulatory DNA is difficult for a variety of technical reasons, for humans, a minimal estimate is that it is one-and-a-half times that for DNA-specifying proteins.

EVOLUTIONARY CONTINGENCY

Why is the human genome so odd? Biologists often try to explain features of living organisms by pointing to their good design and arguing that they must have been the result of evolution through natural selection. When such an explanation is possible, the feature is shown to be an *adaptation*. But it is hard, if at all possible, to see the human genome's architecture as an adaptation. Michael Lynch, a prominent evolutionary biologist from Indiana

University, has been arguing for decades that natural selection has had little to do with the strangeness of the human genome. It is a result of the physical propensities of the DNA of which genomes are made. Pieces of DNA opportunistically multiply themselves, move around, and insert themselves wherever they can. Much of this is probably slightly harmful to the organism. But this harm is not enough for the changes to be discarded and removed from populations through natural selection. All large animals have comparatively small populations and, in such small populations, natural selection is simply not strong enough to remove bloated chromosomes with abundant useless DNA. Lynch points out that bloated genomes are not found in small organisms which typically have large populations in which natural selection can effectively remove even slightly harmful variants. The architecture of the human genome—and other genomes—is largely the result of physical laws, on chance and contingency from a biological point of view.

More traditional evolutionary biologists have tried to answer the apparently inexorable logic of Lynch's argument though none of these responses have been particularly convincing. For instance, if genes are clustered, then breaks in chromosomes which often occur randomly during the formation of gametes, are much more likely to take place in a long nonfunctional segment than in a tiny functional one. Thus, such a structure could be the result of selection. The problem is that there is an equally plausible counter-argument: if a break, however rare, does occur in a functional part in such a structure, the results may be much more devastating. Arguments of this sort can go back and forth endlessly and they can rarely be settled using data that can be collected. (After all, we are talking about events that took place long ago in evolutionary history.) The dust from Lynch's arguments will take time to settle. Meanwhile, the evolution of genomic architecture will remain one of biology's most intriguing problems. Here, at least, the HGP has changed the course of biological research.

In 1991, in an enthusiastic endorsement of the fledgling HGP, Collins had argued that the HGP "will yield a harvest of information that will drive the research enterprise for at least the next 100 years." Collins may well be right though not for any of the reasons he gave at the time. It may well take until the end of the century to understand why the human genome—and other genomes—have evolved to have so strange an architecture. No such positive assessment is possible for the medical effects of the HGP. But, perhaps, that will change with CRISPR. We turn to the creation of that technology next.

Chapter 4

The CRISPR Revolution

> "Tomatoes that can sit in the pantry slowly ripening for months without rotting. Plants that can better weather climate change. Mosquitoes that are unable to transmit malaria. Ultra-muscular dogs that make fearsome partners for police and soldiers. Cows that no longer grow horns. These organisms might sound far-fetched, but in fact, they already exist, thanks to gene editing. And they're only the beginning. As I write this, the world around us is being revolutionized by CRISPR, whether we're ready for it or not."
>
> —Jennifer A. Doudna and Sam Steinberg, 2017, *A Crack in Creation.*

THE CRISPR STRUCTURE EMERGES

The CRISPR story begins a generation ago in Osaka, Japan, in the laboratory of Atsuo Nakata at the Research Institute of Microbial Diseases of Osaka University. In this laboratory in the mid-1980s, a beginning postdoctoral researcher, Yoshizumi Ishino, was sequencing the *Escherichia coli iap* gene that encoded an enzyme involved in alkaline phosphatase metabolism, an important process for breaking down complex molecules into simpler ones. (The acronym *iap* comes from isozyme of alkaline phosphatase. An isozyme is a type of enzyme.) Ishino was working with the K–12 strain of *E. coli* that had emerged as the workhorse of molecular genetics since the 1960s. To understand the gene's functioning better, he sequenced the DNA regions flanking it on both sides. The hope was to find regulatory sequences in these flanking regions as are often seen for bacterial genes such as those of the *lac* operon. These regulatory sequences are DNA segments to which other molecules can attach to turn a gene on or off.

However, instead of a regulatory sequence, Ishino and his collaborators noticed an unusual pattern in the flanking region downstream from the gene: five almost-identical repeated segments each consisting of twenty-nine DNA bases, and separated from each other by thirty-two bases of variable DNA (that came to be called "spacers"). Moreover, the repeated sequence was partially palindromic:

CGGTTTATCCCCGCT**CGCGGGGAACTC,

where the underlining indicates the seven-base palindromic part (which is palindromic by base pair complementarity, that is, A:T and C:G) and the "*" indicates that there was some variability (either G or A was present) between the five repeats.

While many types of repeated DNA sequences were already known in bacteria—and, indeed, in all genomes that had been studied at the sequence level by the mid-1980s—this structure was novel and strange enough to be worth mentioning in the published paper. After noting its presence, Ishino and his coauthors declined to speculate on any function that it could have. They simply concluded: "So far, no sequence homologous to these have been found elsewhere in prokaryotes, and the biological significance of these sequences is not known."

Subsequent work by Nakata's group searched for this structure in a wide array of pathogens. They found it in two other strains of *E. coli* as well as in two other bacterial species, *Shigella dysenteriae*, which causes dysentery in humans, and *Salmonella typhimurium*, which causes gastroenteritis. They did not find it in a few other bacterial species, for example, *Klebsiella pneumoniae*, a cause of pneumonia, or *Pseudomonas aeruginosa*, another cause of pneumonia and multiple other illnesses. In the early 1990s, the structure was also found in a distant bacterial species, *Mycobacterium tuberculosis*, which causes tuberculosis, but not in the closely related *Mycobacterium leprae*, which causes leprosy. The structure seemed to be fairly common, at least in pathogenic bacteria, but not universal. Many bacterial species did not have it. No one had any good idea why it was there.

Interest in these structures or arrays increased after Francisco Mojica, then a graduate student at the University of Alicante, and collaborators began reporting their presence in the archaea, first in *Haloferax mediterranei* in 1993 and then in *Haloferax volcanii* in 1995. Mojica continued to look for these structures in other species during his years as a graduate student and, later, as a faculty member at Alicante after a short postdoctoral stint at Oxford. By 2000, he had found these structures in twenty-five different microbial species.

There was no longer any doubt that the occurrence of these structures was a phenomenon that begged exploration and explanation.

Progress was slow at the time because genomic DNA sequences were rare; indeed, most initial reports of these peculiar repeats came from sequences inferred indirectly through what are known as hybridization studies. That situation changed in the late 1990s with the advent of rapid DNA sequencing techniques and bioinformatics software tools for exploring DNA sequences, all spawned by the HGP. A large library of sequences, particularly of prokaryotes, rapidly accumulated and these arrays were found to be ubiquitous in archaea and very common in bacteria. We now know them to be present in about 90 percent of the former and over 40 percent of the latter. These structures have never been found in eukaryotes.

ITS FUNCTION IS DECODED

In the late 1990s and early 2000s, various names and acronyms were proposed for these odd structures. These include direct variable repeats (DVR), tandem repeats (TREP), long tandemly repeated repetitive sequences (LTRR), short regularly spaced repeats (SRSR), large clusters of tandem repeats (LCTR), and spacer interspersed direct repeats (SPIDR). The one that eventually stuck was coined in 1992 in one of the first bioinformatics-based study of these sequences. Ruud Jansen and his collaborators proposed the name CRISPR for clustered regularly interspaced short palindromic repeats. They argued that the acronym accurately "reflects the characteristic feature of family" and world seems to have agreed with them.

Four features are distinctive of CRISPR arrays: they occur between genes; they contain multiple short repeats with very little sequence variation; these repeats are interspersed with "spacer" sequences that differ from each other; and there is a common leader sequence of several hundred bases on one side of the array. The same paper that introduced "CRISPR" also noted that these arrays occurred close to a family of genes that had been hypothesized to be a "DNA repair system"; it called these CRISPR-associated or *cas* genes (encoding Cas proteins).

Today we know that the number of CRISPR arrays in a genome can vary between two to several hundred. Each CRISPR repeat consists of 25–35 base pairs, varying between species and between different arrays within the genome of each species. Similarly, the spacers are between thirty and forty base pairs long. The number of repeats also varies and can increase, as we shall see, with the addition of more spacers by a single prokaryotic cell. Each CRISPR array is preceded by a "leader" sequence flanking that are the *cas* genes. There is a bewildering variety of Cas proteins and it took a long time to figure out each of their functions and to classify all CRISPR systems into two major types, each with multiple sub-types.

It also took a while to figure out what the function of CRISPR systems are as a whole, and how it is brought about at the molecular level. In 2005, three groups independently recognized that the CRISPR spacer sequences were identical or very similar to DNA sequences from bacteriophages, archaeal viruses, and other pathogens that preyed on prokaryotes. One of these groups was that of Mojica at Alicante. Mojica had been systematically searching for these similarities using nascent bioinformatics techniques. He finally struck success with an *E. coli* spacer that matched the sequence of a P1 phage virus that was capable of infecting many *E. coli* strains but not the one he was working with. By the time they were finished, the Alicante group had found eighty-eight spacers that had matches to known DNA sequences; 65 percent of these were from bacteriophages and plasmids and must have originated from them. They also observed that bacteria were immune to phages for which they had very similar spacer sequences while the bacteria without such sequences remained susceptible to infection.

This was the first indication that the CRISPR system was involved in bacterial immunity. Though now recognized as a seminal contribution, Mojica's difficulties with publishing these results are a matter of legend. The paper was rejected in 2003 by *Nature* with the editor claiming incredibly that the key idea was well known; by the *Proceedings of National Academy of Science (USA)*, *Molecular Microbiology*, and *Nucleic Acid Research* in 2004 (with the first of these journals finding the results lacking sufficient "novelty and importance"), before finally appearing in the *Journal of Molecular Evolution* in 2005 after twelve months of review and revision. Given how CRISPR has since become an icon of contemporary molecular biology there is some irony in this story.

The other two groups were from France, new to CRISPR research, and not focused on the system itself rather than on how it could potentially be used for other purposes. The first of these groups was based in the French Ministry of Defense and also experienced publication difficulties similar to Mojica. Their focus of interest was *Yersinia pestis*, the bacterial agent of plague. In the late 1990s, faced with intelligence reports that the Saddam Hussein regime in Iraq was developing biological weapons, the Ministry wanted the creation of techniques to detect subtle differences between strains of disease agents that could be used to identify the source of their origin. Christine Pourcel from that group thought that the CRISPR locus was a promising candidate because it varied between strains of the same prokaryotic species: different strains had different spacer sequences at the beginning of the CRISPR array. The group noticed that many of these spacer sequences corresponded to sequences of a prophage (that is, the DNA sequence of a phage virus) present in the *Y. pestis* genome and the group prophetically concluded: "CRISPRs may represent a memory of 'past genetic aggressions'." This paper was the first to suggest that prokaryotes with CRISPR arrays were absorbing foreign DNA. *Proceedings of National Academy of Science*, *Journal of Bacteriology*, *Nucleic Acid Research*, and *Genome Research* rejected the paper before it appeared in *Microbiology*. At the French National Institute for Agricultural Research, Alexander Bolotin and his collaborators published a third paper in 2005 pointing out the foreign origin of spacer sequences. This paper was comparatively well received by reviewers: it was rejected only once before appearing in *Microbiology*.

The next sequence of developments followed rapidly. In 2006, Kira Makarova, Eugene Koonin, and their collaborators at the US National Center for Biotechnology Information systematically developed the idea that CRISPR is a prokaryotic immune response system. Much of the theoretical work on the fundamental biology of CRISPR continues to come from that group. Its work, like that of Jansen, made extensive use of bioinformatics techniques, particularly the comparison and analysis of DNA sequences. Techniques spawned by the HGP were finally beginning to make their mark on contemporary molecular biology beyond the small circles of bioinformatics aficionados also spawned by the HGP.

Perhaps even more importantly, experimental confirmation of the immunity hypothesis came soon afterward. In 2007, a team at the Danish food ingredient company, Danisco, provided that confirmation with very convincing data. The team included Philippe Horvath and Rodolphe Barrangou, both scientists employed by Danisco as well as Sylvain Moineau, a phage biologist from Université Laval in Québec, who had been collaborating with Danisco biologists since 2000. This team had long been interested in the yogurt bacterium *Streptococcus thermophilus* because of its commercial importance for the food industry. Part of their goal was to develop DNA-based techniques for the precise identification of bacterial strains as well as to overcome the frequent phage viral infections that plagued yogurt making. Horvath had been using CRISPR sequences for strain differentiation since 2002 and the team was well aware of the CRISPR-based immunity hypothesis which they then proceeded to test.

Working with a well-characterized phage-sensitive (that is, susceptible) strain of *S. thermophilus*, they first showed that exposure to phage led to the incorporation of phage DNA into the CRISPR spacers of the bacterial cells. They then went on to add and delete spacers and showed that the sensitivity of the bacteria to infection by the virus depended on spacer composition: as we should expect, if there were more copies of the phage DNA in a cell's CRISPR array, the greater its resistance to infection by the virus. Moreover, they found that bacteria that had lost their phage resistance had mutant CRISPR spacer sequences that were no longer identical to the phage DNA sequence. They also showed that *cas* genes were involved in both the acquisition of phage DNA and in "interference," that is, resistance to phage infection and multiplication. In particular, they were the first to recognize the function of Cas9 protein as a necessary component of resistance in the species they were working with.

The publication of this paper can be viewed as completing the picture of CRISPR as a prokaryotic immune system though many details still needed to be worked out. The years between 2008 and 2010 saw a flurry of results. At Northwestern University in Evanston, near Chicago, Luciano Marraffini and Erik Sontheimer showed that CRISPR targets DNA rather than RNA as had generally been believed up till time. In Amsterdam, John van der Oost, working with many collaborators including Koonin and Makarova, worked out the details of how the entire CRISPR array is first transcribed into a long single RNA molecule and then processed into smaller functional units. Also, in 2008, Barrangou, Horvath, and their collaborators introduced the term "protospacer" to refer to the DNA segment that was the source of a spacer and is later targeted by it. In 2009, the Alicante group showed that CRISPR interference using Cas9 required the presence of a conserved motif (short DNA sequence, typically two-to-four base long) next to the DNA targeted for acquisition. They called these protospacer adjacent motifs (PAMs). In 2010, Moineau and several collaborators, including the Danisco team, showed that viral DNA targeted by the CRISPR system was sliced apart within the sequence that was complementary to the CRISPR spacer RNA sequence.

THE MECHANISM OF IMMUNITY

The picture of CRISPR-based immunity that has emerged in the past decade consists of an elegant mechanism though some of its details still remain to be worked out fully. Put very simply, what happens is the following: the spacer sequence of CRISPR arrays in prokaryotes are derived from the DNA sequences (or protospacers) of invading pathogens and thus constitutes an immune memory. When there is a new invasion by the same pathogen, RNA transcribed from the spacer recognizes the corresponding protospacer sequence in the pathogen by base pair complementarity (A:T and C:G). Accompanying Cas proteins then cut the DNA and incapacitate the pathogen. This entire process can be usefully divided into three stages: *adaptation*, *expression* (or *processing*), and *interference*. We will consider each in some detail (see figure 4.1).

Adaptation is the process of acquiring a spacer from an invading pathogen. In the most common adaptation strategy, the process begins with a complex of Cas proteins binding to the targeted DNA. This complex then migrates down the DNA until it finds a specific two-to-four base PAM or protospacer adjacent motif. (These PAMs are critical to the functioning of

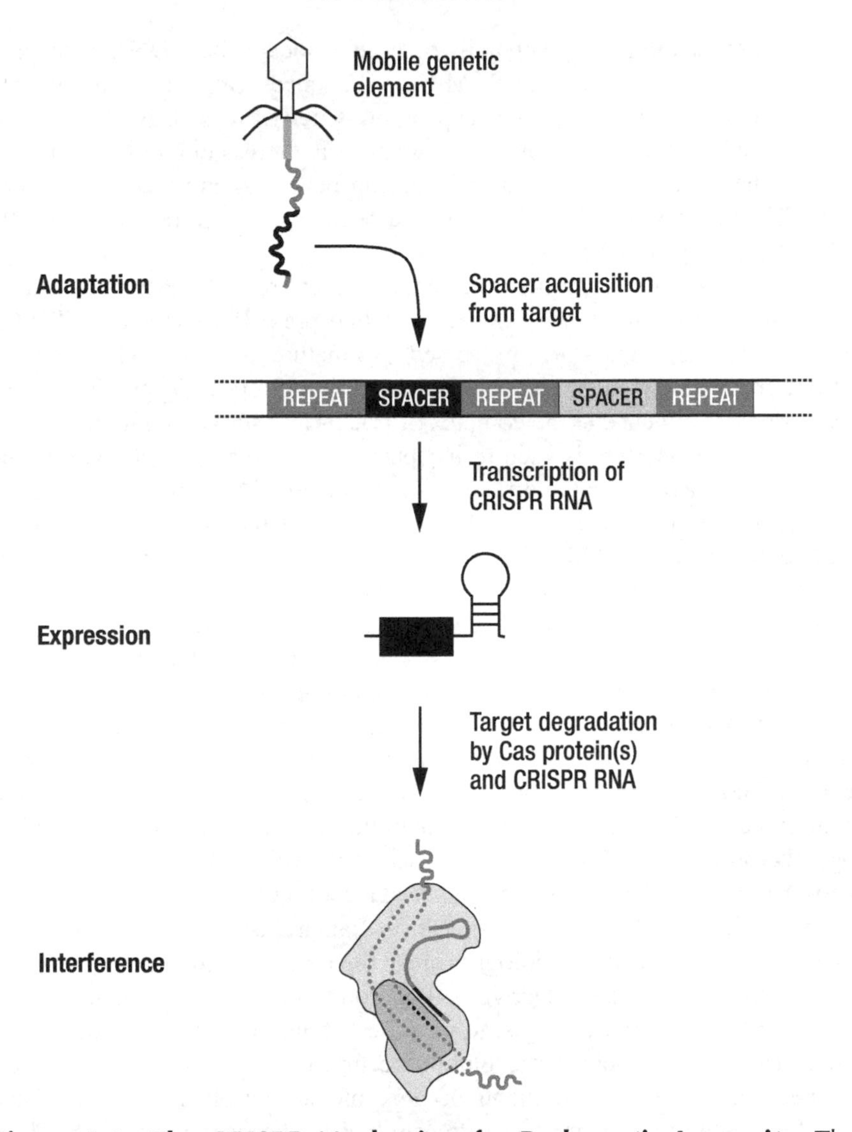

Figure 4.1 The CRISPR Mechanism for Prokaryotic Immunity. The mechanism has three stages as explained in the text. During the first adaptation stage, the prokaryotic cell acquires a spacer from an invading pathogen. During the second expression stage, the CRISPR DNA sequence is transcribed and processed into CRISPR RNA (crRNA) sequences. In the third interference stage, the crRNA binds to a processing complex containing Cas9 (or other protein that cuts DNA) and latches on to the DNA of an invading pathogen which is then cut by Cas9.

the Cas proteins that will eventually be used to incapacitate DNA sequences in the interference stage. The PAM base sequences vary between bacterial species and, within each species, depend on which of these Cas proteins will be involved.) Once the complex finds a PAM, it cleaves out an adjacent portion of the targeted DNA from the invading pathogen and then inserts it in the CRISPR array of the host prokaryotic cell between two repeats, typically at the beginning of the array.

In the *expression* or *processing* stage, the entire CRISPR array is first transcribed into a single long transcript called pre-crRNA (for pre-CRISPR RNA). This transcript is then processed into mature crRNAs each consisting of one transcribed spacer and part of the adjoining repeat. Typically, one of these remains attached to the complex of Cas proteins involved in processing and this is the one that is used to incapacitate the invading pathogen in the *interference* stage. In that interference stage, the crRNA bound to the processing complex is used as a guide to find a region of the DNA in an invading pathogen next to a PAM. The PAM is required for Cas proteins (such as Cas9) to latch on and cleave the DNA. When invading DNA is then cleaved and deactivated the prokaryotic cell has fought off the invasion.

Once a prokaryotic cell acquires a spacer, it will be inherited by its descendants. Thus the memory of the invasions continues down the lineage. Since the presence of a spacer obviously increases the fitness of bacteria in an environment in which the corresponding pathogen with the corresponding is present, CRISPR-based immunity appears to be a case of an inherited acquired adaptive trait. This possibility interests biologists (and philosophers of biology) because it violates the received view of evolution (sometimes called neo-Darwinism) that denies the inheritance of acquired traits, especially adaptive traits. Such inheritance smacks of Lamarckism, a theory of heredity that has been anathema to biologists since the early twentieth century.

Jean-Baptiste Lamarck, Darwin's illustrious early nineteenth-century predecessor, was the first biologist to propose a general theory of evolution in 1809. However, he held that evolution occurred when animals became better adapted to their environments because of internal physical mechanisms and these adaptive features were transmitted to the next generation. Thus, Lamarck endorsed the inheritance of acquired characters but, contrary to popular belief, the inheritance of acquired characters itself is not distinctive about Lamarckism: almost all early nineteenth-century biologists accepted that idea. Our received view of evolution denies both the existence of internal adaptive mechanisms and the inheritance of acquired characters.

According to the received view, all variation in organisms occurs due to random factors, what we now call mutations. Adaptation occurs if, by chance, a mutation turns out to be beneficial. In that case, natural selection ensures its spread in the population. Twentieth-century biologists believed that experimental results ruled out the Lamarckian mode of evolution. Nevertheless,

Lamarckism became part of the official ideology of the Soviet Union during the Stalin era after the dictator came to believe the claims of a fraudulent plant breeder by the name of Trofim Lysenko who promised to revolutionize Soviet agriculture and dramatically increase food production. Genetics was suppressed and geneticists banned and sometimes imprisoned by authorities. Little wonder that the international community of biologists came to associate Lamarckism with terror and fraud. So, when CRISPR suggests that there may be some validity to Lamarckian evolution, the intellectual—and ideological—stakes are high.

Given this background, it will pay to look at CRISPR-mediated prokaryotic immunity process a little more carefully even if it is only peripheral to the use of CRISPR for gene editing. The first question to ask is whether a protozoan cell, say, a bacterium that is invaded by a pathogen, say, a virus can respond fast enough to mount a response that includes the adaptation stage before it succumbs to the pathogen. Experimentally, we do not know the answer to this question but it does not seem likely that there would be sufficient time. Viruses successfully invade bacterial cells very rapidly. So, how is CRISPR-based immunity supposed to work?

Given the absence of clear experimental results what follows will be partly, but not entirely, speculative. We do have the following results. Any population of viruses will have a small fraction of defective ones that are incapable of reproduction within a bacterial cell after invasion. There is one experiment that has shown that bacterial cells seem to acquire spacers from viruses at a rate that is directly proportional to the number of defective phages to which they have been exposed. More recently, it has also been shown that spacer acquisition in *E. coli* bacteria requires active replication of the protospacer-containing viral DNA within the bacterial cell and that spacers are mainly acquired by the bacteria when viral replication is stalled. These results seem to implicate defective viruses as sources of spacer sequences.

They suggest the following scenario. Since the results are only from bacteria and their viruses, we will confine our explicit attention to them though the same story is likely to be true of archaea and for responses to other pathogens. A population of bacteria gets invaded by a population of viruses. Whenever an intact virus invades a bacterium, that bacterium is killed when the virus starts reproducing within it. Most of the bacteria in an invaded population thus perish during the invasion. However, a small fraction of the bacteria gets invaded by a defective virus, say, one in which replication is stalled. Such a bacterium acquires a spacer and thus becomes immune to future invasions by the same virus. These bacteria are very lucky.

If such a bacterium experiences a new invasion by the same virus (in the same generation, that is, before it splits into two daughter cells), then it has an acquired adaptation that will enable it to survive. Much more likely, it will already have reproduced before experiencing another such invasion. Because the spacer is incorporated into its genome, this capacity is transmitted to its descendants. Thus, the ensuing bacterial lineage has an inherited acquired adaptive trait though luck has played a major role in this process in the sense that the cell that first acquired the spacer was very lucky to have been invaded by a defective virus. Thus, CRISPR does challenge the received view of evolution to some extent. Much more can be said on this topic but it is beyond the scope of this book. We return to the story of how CRISPR, a prokaryotic immune system, led to the best universal gene-editing tool.

ONWARD TO GENE EDITING

Three papers from 2011 set the stage for the emergence of CRISPR-based gene editing. Two of the papers were experimental; the third was a bioinformatics-based theoretical analysis by Koonin and Makarova with collaborators around the world from across the CRISPR research spectrum. The first experimental paper came from Emmanuelle Charpentier's laboratory in Umeå, Sweden. Working with the pathogenic bacterium, *Streptococcus pyogenes*, Charpentier and her collaborators reported a new step during the processing of pre- crRNA into crRNA. This step required the participation of a piece of RNA coded from a different gene, that is a trans-activating crRNA (tracrRNA) that had a 24-nucleotide base sequence complementary to the repeat segments of the crRNAs. So, it now became clear that three types of molecules were necessary for interference, crRNA, tracrRNA, and Cas proteins besides the usual cellular machinery of the bacteria (and, presumably, the archaea).

The second experimental paper, from the laboratory of Virginijus Siksnys at Vuknius University in Lithuania, was crucial to the emergence of a CRISPR-based gene editing. This paper reported two crucial results. The collaborators (among whom were the two Danisco biologists, Barrangou and Horvath) managed to transfer DNA containing a CRISPR locus from one species to a distant one. They showed that the CRISPR mechanism continued to function appropriately in the new species. The transfer was from *S. thermophilus* (which explains the interest of Danisco) to the laboratory, workhorse, *E. coli*. The significance of this result was that it raised the possibility that a CRISPR-based system could be transferred to new contexts without loss of function. The paper also showed that a single Cas protein, now called Cas9, is sufficient to cnsure the proper performance of the CRISPR system. This observation strongly suggested that it could turn out to be relatively simple to construct a gene-editing technique based on CRISPR. This paper also began an analysis of the roles played by different parts (or "domains") of the Cas9 protein in cleaving DNA.

Experimentally, these two papers set the stage for the construction of a CRISPR-based gene-editing technology. The third important paper—the theoretical one—played a catalytic role. This paper was primarily a review of all work on the CRISPR system up to that time; with Makarova and Koonin playing the most prominent roles, the list of authors included almost the entire spectrum of prominent CRISPR researchers. The main achievement of the paper was a classification of the different types of CRISPR systems that had emerged during the past decade. But the paper also summarized what had so far been discovered about the role of the different types of RNA and the role of Cas9, especially in cleaving DNA.

The stage was set for a reconstruction of the CRISPR system as a universal gene (and genome) editing technology that could be used within any species to target almost any DNA sequence (see figure 4.2). Charpentier began a collaboration with Jennifer Doudna, a prominent RNA biologist, at the University of California at Berkeley. In 2012, they published the paper that established how the CRISPR system constitutes a fully programmable gene-editing technology. The team reconstructed a breathtakingly simple version of the CRISPR system *in vitro* (that is, in a test tube). This version consisted of Cas9 protein and a single "guide"

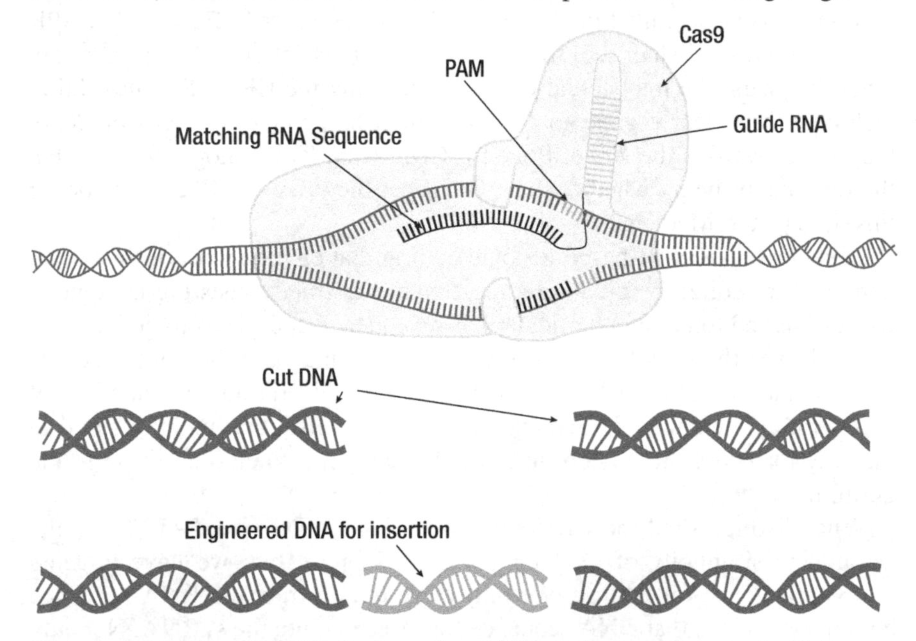

Figure 4.2 The CRISPR System for Editing Genes. The guide RNA recognizes a matching sequence in a gene being targeted for editing. So long as an appropriate PAM is present, the Cas9 enzyme cuts the DNA. Repair is guided by a complex that includes an engineered DNA to be inserted at the site of the cut.

RNA (sgRNA) that was a chimera of the crRNA and a tracrRNA. Any DNA sequence could be targeted for cleavage by programming the spacer sequence of the crRNA segment. The system's competence was demonstrated by constructing a chimeric sgRNA that targeted the gene for the green fluorescent protein. They showed that different domains of the Cas9 molecule cleaved the two segments of the targeted DNA. The collaborators were fully aware of what they had achieved. After noting the recent successes of ZFNs and TALENs in manipulating genomes, the paper concluded by proposing "an alternative methodology based on RNA-programmed Cas9 that could offer considerable potential for gene-targeting and genome-editing applications."

Shortly afterward, in February 2014, a group from the Broad Institute, jointly hosted by Harvard University and the Massachusetts Institute of Technology, reported using the CRISPR technique to edit the genes of eukaryotic cells. George Church and his group at the Harvard Medical School reported using the technique to edit genes in human cells. Doudna and Charpentier were correct: for the first time, we had a universal programmable gene-editing tool. Subsequently, a legal patent fight over the technology erupted pitting the Broad Institute against Doudna and Charpentier. The former has largely prevailed in the United States; the latter in Europe. CRISPR is so useful (as we shall see) that billions of dollars (or euros) were at stake. Over the years, the intellectual credit for devising the CRISPR gene-editing technique has clearly gone to Doudna and Charpentier: in October 2020, they were awarded the Nobel Prize for Chemistry. The honor was expected, the only doubt being whether they would get the prize for Chemistry or for Physiology and Medicine.

In standard CRISPR-based genome editing that emerged out of this work, genomes are edited by creating a molecular construct consisting of a guide RNA designed to target a particular DNA sequence and genes to be inserted. Cas9 cleaves the targeted DNA in the chromosomes and the construct gets inserted into it. This technique enables the rapid editing of somatic cell and germline genomes. It was realized from the onset that nothing stood in the way for using this system to target human genes including those in the germline.

What distinguished the CRISPR system from ZFNs and TALENs is the astounding simplicity of its programming. Suppose that we have targeted a DNA sequence for editing. We only have to construct the RNA sequence complementary to that DNA sequence for insertion into the sgRNA. Not only is this conceptually simpler than engineering protein segments in the case of ZFNs and TALENs, it is experimentally much easier—and much much cheaper—to do. Little wonder that the CRISPR system immediately replaced its predecessors completely. Once the target DNA is cleaved at the desired site, DNA engineered for insertion can be inserted at that site by having been part of what was put into the sg RNA. For instance, a functional version of a gene can be inserted to replace a dysfunctional one.

The standard CRISPR system uses Cas9 to cleave DNA though other nucleases have also been used and may eventually replace Cas9. The Cas9 molecule needs a specific PAM sequence upstream from the spacer to cleave DNA. The most commonly used Cas9 molecule comes from the bacterial species, *Streptococcus pyogenes*, and its PAM is NGG, where "N" means any nucleotide from the standard quadruplet (A, T, C, or G). What if this triplet is not present just upstream from the targeted protospacer? Cas9 from *S. pyogenes* would not work. But this problem did not turn out to be insurmountable,

Over the years, many ways have been worked out to get around the problem. The simplest is to exploit the natural variability of Cas9 that differs from one species to the next. These different versions of Cas9 require different PAMs. For instance, Cas9 from *S. aureus* uses the PAM sequence NGRRT or NGRRN, where "N" means any nucleotide base and "R" means A or G; Cas9 from the bacterium *Treponema denticola* (which causes human dental disease) uses the PAM sequence NAAN. Some thirty odd variants of Cas9 are in use in laboratories today and the number keeps on increasing. Moreover, nucleases other than Cas9 use other PAMs and there are also strategies to alter nucleases to fit the corresponding region near any targeted DNA space though this is somewhat more difficult. The upshot is that the CRISPR system is truly universally programmable.

But problems remain. The CRISPR system's specificity to its target, that is, whether it edits the intended target DNA sequence *and nothing else*, is not perfect and off-target mutations remain an ever-present danger. While there is consensus amongst CRISPR researchers that this problem is under control and the accuracy of CRISP-based gene editing has continued to increase, in each potential therapeutic case, experimental demonstration of accuracy in laboratory cell lines remains a must before editing can reasonably be tried on live patients, even if it is only somatic cell gene editing. We will have more to say about the problem of off-target mutations induced by CRISPR-based gene editing in the sixth chapter.

THE DELIVERY PROBLEM

Like all other gene-editing systems, the CRISPR system has to be delivered directly to the targeted cells to be effective. When germline genes are targeted, delivery is relatively straightforward. The CRISPR construct, that is, sgRNA and Cas9, can be injected into the single-celled embryo or zygote using a microscale needle. This method has been used in humans with some success to repair point mutations. In these experiments, the Cas9 protein was injected into the cell along with the sgRNA. But Cas9 RNA can also be introduced into the cytoplasm for translation into protein at the ribosomes; alternatively, Cas9 DNA can be introduced into the nucleus for transcription and translation.

The advantage of introducing the Cas9 protein itself into any cell is that the CRISPR system begins to perform its function immediately and leads to rapid gene editing. However, there are two disadvantages. The effect is transient; it lasts only for as long as the protein is not degraded by everyday chemical interactions within the cell. This problem also manifests itself when Cas9 RNA rather than DNA is introduced into a cell. Moreover, direct introduction of Cas9 into a cell is somewhat difficult because Cas9 is a very large protein molecule, about 160 kilodalton in size. This is almost four times the size of a hemoglobin macromolecule with its four chains and heme group. Cas9 is huge.

The problem of introducing Cas9 becomes particularly acute when the goal is to deliver the CRISPR system to specific organs for somatic cell gene editing. The system has to be guided to the right organ and then thrust into the targeted cell across the cell membrane. Not only is Cas9 large but the phosphate backbone of the sgRNA is negatively charged. These are exactly the type of molecules that cell membranes are designed to prevent entering the cell.

Because of the size problem associated with Cas9, there has been a lot of work in designing CRISPR gene-editing systems in which a different nuclease is paired with the sgRNA. One that emerged in 2015 is now known as Cas12a (and was formerly called Cpf1). This molecule is smaller than Cas9 and, in some ways, more flexible. Cas12a has a different PAM than Cas9 and can also be used for those relatively rare genes for which a suitable PAM cannot be found for the Cas9 family. Cas12a also cuts the DNA differently from Cas9 and in such a way that it may lead to even more accurate editing.

For somatic cell gene therapy, delivering the CRISPR system to the targeted organ and into the appropriate cells remains the most glaring technological problem for using CRISPR-based techniques even though a bewildering variety of delivery methods have been developed. Not only does the cell membrane stand in the way, the presence of the sgRNA and Cas9 triggers an immune response in humans. That response is designed to destroy such foreign molecules and, very often, does so quite efficiently.

One standard response to these problems is to use viruses known to be capable of reaching the targeted organs. (We know this because we find these viruses in these organs.) The workhorse for this strategy are the tiny adeno-associated viruses, or adenoviruses, that are known to infect humans but not cause any disease and induce only a mild immunological response. These viruses come in a variety of types each of which infects a different human cell type thus allowing specific organs to be targeted. The use of this virus for therapeutic gene editing in humans has recently received regulatory approval in both Europe and the United States. One problem of using such a tiny virus is that the payload of DNA that it can carry as DNA is small. At least in laboratory studies, it remains the most common vector for the delivery of CRISPR-Cas9 systems into human cells.

Other delivery techniques include electroporation which uses a strong electric field to disrupt the cell membrane surface so as to facilitate entry of the sgRNA and Cas9 payload. This method works in the laboratory for zygotes but is not a promising approach to *in vivo* somatic cell gene editing. Alternatively, high-pressure injection of liquid containing the payload can be used to get the payload into the targeted organ and distributed around the tissue. However, it does not get it into the cells and this method must be coupled with some other that will enable entry of the CRISPR system into the cell and its nucleus. In general, these two physical methods all work reasonably well in the laboratory but are of less value when faced with a living organism.

Chemical methods fare better. The payload can be encapsulated in a lipid nanoparticle that protects it from degradation by enzymes. If these nanoparticles are delivered to the appropriate organ, they interact with the cell membranes, which also consist of lipids, and deliver the payload. Gold nanoparticles have also been used for this purpose and have become increasingly popular after a 2017 report showed that they could be used to target a wide variety of cell types in mice. It is possible we will eventually get a gold standard. Meanwhile, viruses remain indispensable.

UBIQUITOUS CRISPR

CRISPR is everywhere, not merely because bacteria are everywhere, but because anyone can buy a CRISPR plasmid (a circular bit of DNA including the desired sequences) from a nonprofit plasmid repository, Addgene, currently for US$ 65 (as of 2020). By 2018, nearly 3,400 laboratories had received CRISPR shipments and the number by now is obviously much higher. Meanwhile, anyone reading these pages is likely to have eaten fruit or vegetables with genomes modified by CRISPR. While genome editing of food plants using ZFNs and TALENs has been around for a while, its significance seems trivial in retrospect compared to what has been achieved using CRISPR.

Non-browning apples created through CRISPR-based genome editing has been in markets for years. This is not a trivial achievement. Food waste, especially in the form of fruits and vegetables that get old and are thrown away by shops and customers, is a major problem worldwide. It is estimated that nearly half of the produce grown in the United States is thrown away. By reducing that waste, CRISPR already contributes to food security, potentially worldwide.

But CRISPR's ambit in agriculture is not limited to waste reduction. Potatoes have had their genomes edited for waxiness. Tomatoes have had their genomes edited to resist heat stress and powdery mildew disease. Very recently, one group has produced bushy tomato plants with bunched fruits like a flower bouquet instead of our usual long vines. They have pointed out how easy it would be to grow these plants in urban settings. If you live in a city, just imagine growing all your tomatoes on your window sill. Mushrooms have had their genomes edited to decrease browning. The US Department of Agriculture has ruled that this process will not be regulated. This decision frees up many more economically important possibilities.

Maize has had its genome edited for drought resistance. Cassava has been edited for resistance to brown streak disease. Flowering times have been altered in soybeans to increase yield. One group has succeeded in increasing yield in rice. CRISPR may well decrease hunger and food shortages at a scale similar to the Green Revolution of the 1960s. Some commentators have suggested that, thanks to the advent of CRISPR it may be possible to double food production by 2050 without excessive conversion of wildland to cropland. This is the goal that, according to some, must be achieved to feed the projected human population in mid-century. Finally, CRISPR-edited plants are non-GMO; they are considered not to be "genetically modified organisms" in the technical sense of not having DNA sequences from other species engineered into their genomes. Thus, the use of CRISPR-based gene editing should not raise the hackles of the anti-GMO movement.

Turning to livestock, CRISPR has already been used to edit genomes of chickens, cattle, and pigs. In chickens, a gene has been disabled to remove a protein from egg whites that sometimes induces allergic reactions in people eating eggs. Genomes of cattle have been edited to increase resistance to tuberculosis. CRISPR has also been used to create cattle that only produce male calves. The motivation for this was that males grow bigger and faster than females and thus produce more beef. A wide variety of modifications have been introduced in pigs, particularly in China. They have had their genomes edited to be leaner. These leaner pigs have lowered risk of mortality and potentially save farmers money. Resistance to a variety of diseases has also been introduced including porcine reproductive and respiratory syndrome virus (PRRSV) disease; moreover, porcine endogenous retroviruses (PERVs) have been removed from pig genomes to make pork safer for humans.

However, as we shall see in the sixth chapter, genome editing in pigs and other animals (and not only through CRISPR) has also let to a suite of unpredicted and untoward consequences that raise questions about how much more regulation, and of what type, these methods should have before they spread even further. As environmentalists and animal ethicists have long and repeatedly pointed out, the food industry does not have a morally acceptable record of treating sentient animals well. Even without human eugenics, CRISPR brings to the forefront a host of ethical questions that we can no longer avoid. But animal ethics is beyond the main thrust of this book; moreover, in this context, CRISPR is not introducing new challenges. It is extending the scope of old ones introduced by prior methods of genome editing.

Chapter 5

Inevitable Eugenics?

> "Eugenics is a word with nasty connotations but an indeterminate meaning. Indeed it often reveals more about its user's attitudes than the policies, practices, intentions, or consequences labelled."
>
> —Diane B. Paul, 1998, "Eugenic anxieties."

EUGENICS IN THE NEWS

Almost any discussion of the potential of using CRISPR-based methods for human gene editing eventually veers into prospects for eugenics. For many of us, even among those who approve of some forms of eugenics, the word itself seems to carry some nebulous connotation of danger, if not outrage. Why it does so should be clear from the history with which this book began. Those who support eugenics typically act as if they are on a rescue mission to rehabilitate a social program soiled by histories of genocide and involuntary sterilization. And those who object to eugenics often feel that they have to offer no argument beyond pointing to those histories. But what *is* eugenics? So far, in this book, we have not tried to answer this question.

We can no longer do so. Before we decide whether eugenics in any shape or form is acceptable or even desirable, or merely inevitable, we must achieve some clarity about what it is supposed to be. That is not a trivial task. As the historian Diane Paul has noted, eugenics "has been variously described as an ideal, as a doctrine, as a science (applied human genetics), as a set of practices (ranging from birth control to euthanasia), and as a social movement." Even before World War II, Paul points out, eugenics was a source of endless controversy, Paul was writing several decades ago. In the intervening years, the word "eugenics" was somewhat rehabilitated in the context of the expected consequences of the HGP and through the efforts of liberal eugenicists promoting parental choice in human germline intervention. Nevertheless, since 2019, controversies about eugenics, what it means, and how we should regard its advocates from a century ago have publicly erupted and degenerated into the theater of the absurd in Britain.

Galton left his personal collections and archives with the University College London (UCL) where he also endowed a chair of Eugenics that became operative on his death in 1911. The endowment was for forty-five thousand pounds (worth about five and a half million pounds today). The first occupant of the chair was Galton's protege, Karl Pearson. Both Galton and Pearson are towering figures in the history of statistics. Indeed, most historians regard Galton as the founder of classical statistics (sometimes also called frequentist statistics) because of his invention of the concept of regression and elaboration of the concept of correlation. Pearson provided the first quantitative account of correlation through a coefficient named after him. It has ever since been the most widely used measure of correlation. Galton also made many other important scientific contributions including the use of fingerprints to identify individuals.

Until last year, UCL had lecture halls named after Galton and Pearson and a building named after the latter to commemorate how he had established the study of both heredity and statistics at the college. Pearson was an important cultural figure in his day, a committed socialist and ardent champion of women's rights and sexual freedom. Even by the standards of their day, Galton and Pearson were also run-of-the-mill scientific racists believing in racial differences in ability on supposedly scientific grounds. They were also the most prominent eugenicists of their age. As proselytizers for eugenics, Galton's and Pearson's legacies are complicated. What must be emphasized first, though, is that they advocated and emphasized *voluntary* eugenics, a choice to reproduce (or not) based on heredity. Pearson on occasion suggested segregation of the "unfit" but neither advocated involuntary sterilization let alone extermination in death camps.

By 2019 protesters at UCL had had enough of Galton, Pearson, and eugenics in their institutions. Most protesters were students but some faculty members such as Joe Cain of the Department of Science Studies were vocally supportive. The campaign equated eugenics to racism without argument. As one protester put it: "Buildings all over campus are named after eugenicists who today we would call white supremacists." Given the global reach of eugenics in the early twentieth century, the identification is misplaced but the protesters were clearly not willing to engage in the critical reflection that should have marked supposed members of an academic community.

History provides no support for the protesters. Even in the decades following Pearson's death in 1936, eugenic sterilization, for instance, in the United States, targeted poor white women along with women of color. Carrie Buck was white. Arguably, class differences have dominated racial differences throughout the history of eugenics. Pearson and Fisher (to whom we will turn next) feared excessive reproduction of the lower classes. Galton advocated Chinese expansion into supposedly inferior Africa but that hardly constitutes an advocacy of *white* supremacy (though there is other evidence that damns him on this count).

Although most eugenicists were also racists in Galton's and Pearson's era, there was an important difference: eugenics privileged some groups not because of appearance alone but because of hereditary factors that came to be identified with genes after 1900. Galton repeatedly emphasized that how, and to what extent, these factors determined traits and how they were distributed across populations was an empirical question on which data must be collected. The contrast with today's white supremacy movements should be obvious.

As we shall see in the seventh chapter, there are pervasive problems of both obvious and subtle racism in contemporary British society and its intellectual ideology. The protesters at UCL, perhaps inadvertently, drew attention away from these deeper problems as the fixated on the naming of structures after long-deceased flawed figures who also were of some very well-deserved scientific repute. But they succeeded in their goals. The Galton Lecture Theatre was renamed Lecture Theatre 115; the Pearson Lecture Theatre became Lecture Theatre G22; and the Pearson Building became the North-West Wing. In January 2021, UCL issued a formal public apology for its history and legacy of eugenics. But UCL made no move to divest itself of its Galtonian endowment. It continues to exploit Galton's largesse. If the protest movement had only targeted racism, there may have been some good reason to support these developments. By targeting eugenics with no concern for detail, it lacked intellectual integrity.

Reveling in its dazzling eminence, the University of Cambridge was not going to be outdone by UCL. There, Caius Hall in Gonville and Caius College used to have a set of six celebratory stained glass windows commemorating the most famous mathematicians and scientists to have worked in the College: Sir Charles Sherrington, the neurophysiologist who has received a Nobel Prize in 1932 for his work on the function of neurons; the mathematician, John Venn, famous for the diagram in set theory named after him; another mathematician, John Green, who proved an important theorem in vector calculus; Francis Crick, whom we have encountered before (and also a Nobel Laureate in 1962); James Chadwick, who won a Nobel Prize in physics in 1935 for discovering the neutron; and Roland Aylmer Fisher,

Pearson's successor at UCL who was probably the greatest statistician of the twentieth century besides being, along with Haldane and Sewall Wright, one of the founders of modern evolutionary theory. Fisher was also an eugenicist.

In 2020, the Fisher window became the target of Black Lives Matter and allied protesters at the University of Cambridge. Protesters proclaimed "eugenics is genocide—Fisher must fall" on the Gonville and Caius College Gate of Honour. (For full disclosure: I am part of the Black Lives Matter movement in the United States). Once again, eugenics was identified with racism by the protesters. That Fisher was a committed eugenicist had always been well known though the College authorities preposterously claimed not to have been "fully" aware of his views when the window was installed in 1989. Apparently none of those responsible was aware that the entire second half of Fisher's 1930 book, *The Genetical Theory of Natural Selection*, one of the principal sources of modern evolutionary theory, was devoted to eugenics. It is also not a matter of dispute that Fisher encouraged voluntary eugenics through economic incentives but did not advocate involuntary sterilization, let alone death camps.

But Fisher's detractors also accused him of racism. Here, the issues are far less clear. Politically, Fisher was an imperialist conservative from Britain's colonial era and it is well-nigh impossible to find anyone of such a political persuasion who was not also a racist. Some of those who knew Fisher—for instance, the recently deceased Raphael Falk, a prominent geneticist and historian of science at the Hebrew University of Jerusalem—thought that he was a racist. Yet, Fisher was a repeat visitor at the Indian Institute of Science in Calcutta (now Kolkata) and is known to have mentored Indian students there beyond the call of duty. He also maintained cordial and supportive relations with black African statisticians who were former students. Of course, being on civil terms with some select persons of color does not prove someone is not a racist. But evidence of racism is absent in Fisher's extensive archives at the University of Adelaide. However, once again, absence of evidence is not necessarily evidence of absence.

Fisher's detractors point to a passage quoting him in a 1952 United Nations Educational, Scientific and Cultural Organization (UNESCO) statement on the "Race Question in Modern Science." Founded after the end of World War II, UNESCO had optimistically taken upon itself the role of scourging the world of scientific racism as had been promoted by Nazi and other Northern scientists. Its first statement on race, from 1950, took the view that race (that is to say, our usual racial categories) had no biological basis and was a social construction that should play no role in biological research. It generated a strong backlash from a relatively small but very vocal cadre of white scientists. A more timid statement followed in 1952.

The new statement did not satisfy Fisher who had also been one of the critics of the original one. The 1952 report notes:

> Sir Ronald Fisher has one fundamental objection to the Statement, which, as he himself says, destroys the very spirit of the whole document. He believes that human groups differ profoundly "in their innate capacity for intellectual and emotional development" and concludes from this that the "practical international problem is that of learning to share the resources of this planet amicably with persons of materially different nature, and that this problem is being obscured by entirely well intentioned efforts to minimize the real differences that exist."

In the dispute at Gonville and Caius College, Fisher's detractors were happy to use the first quotation from him but ignore the second.

But what does the first quotation establish? It is not surprising coming from Fisher. But this is not because of whether or not he was a racist. It is simply because that he, like most other geneticists of his generation, was beguiled by the lure of genetic reductionism (a doctrine we have encountered before). The assumption was that genes pretty much determined even complex human behavioral traits, including the intellectual and emotional capacities mentioned by Fisher. If you believe that, and you are convinced (as Fisher was) that people from within a racial category (as understood at the time) share more genes with each other than with people from other races, you are logically committed to the belief that there are racial differences in these traits.

This is a classic example of what philosophers call a *valid* but *not sound* (that is, *unsound*) argument. It is valid because the conclusion does logically follow from the premises: it cannot be the case that all the premises are true and the conclusion false. It is unsound because some or the premises are false and this means that we have no reason to accept the conclusion. In fact, the two important premises are both false though that was not clear in 1952 when Fisher was making this argument. First, genetic reductionism is simply not a defensible view. Genes alone do not play a determinative role in the origin of complex intellectual and emotional traits. That is one of the most important—and unintended—results generated by the HGP. We have already encountered this insight in the last chapter and will return to it in some detail in the seventh chapter. Second, it has turned out to be false that people from the same race (as usually understood, for instance, Black and white) share more genes with each other than with people from a different race. There is no biological basis for our usual racial categories (but the details for that conclusion are the matter of a different book).

What is more relevant to us is that the Fisher window at Gonville and Caius College was being targeted mainly for Fisher's advocacy of eugenics. To its left (when facing the panel of windows) was the Crick window. The Cambridge protesters did not object to it. So, let us look at what Crick had to say about eugenics. In a 1962 London symposium on "Man and His Future," some ten years after Fisher's infamous remarks, Crick had the following "insight" to contribute:

> Let us take up this whole question of eugenics. I think we would all agree that on a long-term basis we have to do something . . . I want to concentrate on one particular issue: do people have the right to have children at all? It would not be very difficult, as we gathered from [another participant], for a government to put something in our food so that nobody could have children. Then possibly—and this is hypothetical—they could provide another chemical that would reverse the effects of the first, and only people licensed to have children would be given the second chemical. This isn't so wild that we need not discuss it. Is it the general feeling that people have the right to have children? This is taken for granted because it is part of Christian ethics, but in terms of humanist ethics I do not see why people should have the right to have children. I think that if we can get across to people that their children are not entirely their own business and that it is not a private matter, it would be an enormous step forward.

Crick's remarks were well received. Fisher was dead; were he alive he likely would have blanched at the idea of governmental interference into personal reproduction in this way.

Fifteen years later, Crick laid out his eugenic vision more explicitly. He was quoted by the *Pacific News Service* as saying: "No newborn infant should be declared human until it has passed certain tests regarding its genetic endowment and that if it fails these tests, it forfeits the right to live." His remarks became a foil for the emerging raucous anti-abortion movement in the United States.

The Gonville and Caius College authorities have removed the Fisher window. The Crick window remains. Around the same time, Cambridge hired a University lecturer, Staffan Müller-Wille, who is an apologist for Carl Linnaeus. While Linnaeus is mostly known for classifying the living world in the eighteenth century with a binomial scheme that we still use today (where every species has a genus name followed by a specific name, for instance, *Homo sapiens*), he is also the principal progenitor of scientific racism introducing a hierarchy of human races to match the prejudices of his day. Müller-Wille is apparently not interested in whether the concept of race is a "false idea"; rather his concern has been about how the concept can be a useful "mental tool." Postcolonial readers will have no difficulty in recognizing for what purposes race has been such a useful mental tool. Let us end on a positive note: we cannot justly accuse the Cambridge protesters and authorities of consistency and the littleness of mind that goes with it.

WHAT IS EUGENICS?

Eugenics, in spite of these controversies and the scores of volumes written about it, has never been precisely defined successfully. In the first chapter, which discusses the history of eugenics, we made no attempt to formulate precisely what "eugenics" means, leaving that task for later in the book should it become necessary. Let us begin by noting one feature about eugenics that does emerge from the discussion of that chapter: eugenics includes, in some form or other, an attempt to change the genetic future of human populations through conscious and directed intervention in reproduction. But, beyond that, there is little agreement. Sterilization and the murder of those deemed unfit definitely do constitute eugenics. Merely choosing a mate whom one finds attractive almost certainly does not. Marrying for money certainly does not. Using contraceptives falls into the gray area between what may or may not be eugenics. It depends on the motivation. We will shortly have more to say on these issues. Other such intermediate cases are even murkier. When Indian parents demand fair brides for their sons to have light-skinned grandchildren, is that eugenics?

What makes the situation difficult is that the simple formulation proposed in the last paragraph, no matter how initially plausible it may seem, is riddled with problems. Historically, many prominent advocates of eugenics have supported the idea that it be promoted through reproductive decisions such as the use of contraceptives. These advocates included many prominent supporters of birth control in both Britain and the United States since the 1920s. Contraception was viewed by such eugenicists as one of the most appropriate modes of reproductive intervention. Now, suppose you use contraception only because you have no intention of having more children. Is that not a conscious decision to alter the genetic future of your population? What could be objectionable about such a choice? Even if it is eugenics, where is the nasty connotation?

Conversely, when contraception is rejected, the reasons for doing so determine the ethical status of that decision. For instance, you may reject contraception because you believe in the superiority of your own religion and feel obligated to increase the number of its adherents compared to "inferior" religions. (There are such factions within many major religions who come to mind, including extremists from Christian, Hindu, and Islamic traditions.) So, you may find contraception nasty—but the nastiness comes from your religious predilections, not from contraception itself. Yet, if religious persuasion in a context is correlated with the genetic profile of a population because of an association between particular religions and ethnicity, a decision to forgo contraception will make a consciously chosen difference in the genetic profile of future generations. Is this eugenics? On the other side, the environmental problems posed by global overpopulation (and continued rapid population expansion in large swathes of the South), and the resulting overconsumption, makes birth control seem like a moral imperative to some people rather than a problem. Suppose we choose contraception for this reason. Is this eugenics?

Our initial formulation would also make much of genetic counseling eugenic. Let us return to the case of Tay-Sachs disease that we discussed earlier (in the second chapter). There is an elevated presence of the Tay-Sachs allele in Ashkenazi Jewish communities in Canada and the United States. In many such communities, effective genetic testing and counseling was used to decrease disease incidence by 90 percent between 1970 and 2000. Either individuals who were heterozygous for the allele were discouraged from marrying each other or homozygous fetuses were aborted. Objections to abortion aside, what are the negative connotations of this kind of eugenics?

Many accounts of eugenics would require it to include a dose of coercion thus contributing to the nasty connotation that Paul mentions. These accounts would cover cases such as those of the first chapter such as widespread involuntary sterilization in the United States and the Nazi atrocities that generated widespread revulsion toward eugenics. The plausibility of framing eugenics in this way comes from the fact that it would exclude from eugenics measures such as contraception and genetic counseling followed by voluntary non-reproduction.

But, what about some other noncoercive measures such as providing monetary incentives for reproduction by those deemed to be genetically superior? Such payments were promoted by some British eugenicists of the early twentieth century such as Fisher, who were happy to be associated with eugenics but, as we saw earlier, were not willing to stoop to coercion. Isn't there something troubling about the state deciding to encourage only a subset of its populations to breed because of allegedly superior biological inheritance?

Or, take the case of Singapore. Starting in the 1960s, the city-state's population policies were guided by the assumption that desirable human behavioral and mental traits are genetically determined. Singapore's elite, including its political leadership, were happy to promote eugenic policies explicitly using that term. Their intent was not only to reverse the ongoing decline of the total population of Singapore but also to encourage the selective breeding of superior genotypes which were identified primarily on the basis of academic achievement. In the early 1980s, the government unveiled a package of measures to promote mating among highly educated individuals, including

> a computer dating service; fiscal and other incentives for graduate women to bear more children; love-boat cruises (all expenses paid) for eligible graduate singles in the civil service; special admissions criteria to the National University of Singapore (NUS) to even out the male-female student ratio; calls to NUS academicians to investigate the single graduate problem, and also the introduction of courtship classes in the undergraduate curriculum to hone the would-be suitor's skills, etc.

Coercion had happily lost out to absurdity in Singapore.

DEFINING EUGENICS

Given these problems, is there any point in trying to define eugenics? It will obviously not be easy and almost certainly not fully successful. Nevertheless, we will try to give a working definition because some clarity will be needed when we examine programs and strategies that are either criticized or promoted as eugenic. These have proliferated since the 1990s, encouraged by the HGP, and now spurred on by CRISPR-based potentialities. The first issue to address is one that was raised by Paul on which we have commented already: Why eugenics carries a nasty connotation for most of us (though apparently not for the leaders of Singapore and like-minded people), or at least did so before recent attempts began to rehabilitate the term?

The crucial reason is the obvious one: the historical connection between eugenics and grotesque forms of coercion. Tremendous harm has been done in the name of eugenics (to use Dan Kevles's phrase) and it is hard to get away from images of Nazi death camps and young destitute women being sterilized in the United States. If this is what eugenics means in practice, little wonder that most of us want no truck with it. But contrast this picture with the Tay-Sachs story and think of the possibility that, with the technologies we now have, it can be entirely eliminated along with many other genetic diseases. Do we want to prevent such measures?

In what follows an important distinction will be that between positive and negative eugenics. This distinction goes back to the early twentieth century, to a British physician named Caleb Williams Saleeby. In 1909, in *Parenthood and Race Culture*, Saleeby distinguished between two forms of eugenics, "one [that] would encourage parenthood of the worthy, the other discourage parenthood of the unworthy." This distinction became a staple of eugenics, especially in Britain. It was particularly relevant to liberals such as Haldane who endorsed some forms of eugenics in the 1920s.

By and large, these liberals were much more enthused by the prospects for negative eugenics, especially the elimination of genetic diseases than in trying to increase the prevalence of desired traits in a population. Eliminating single dominant genes that caused disease was fine for them. Some liberals were also willing to eliminate recessive genes that caused disease though, as Haldane soon realized and pointed out, this would be much more difficult. Trying to select genes supposedly linked to desirable complex traits such as intelligence or those related to temperament was a different matter altogether.

But even here the advent of the Nazis transformed the landscape. Haldane, for instance, had been sympathetic to some forms of eugenics into the early 1930s; by 1938, as we saw in the first chapter, he was expressing skepticism about any use of heredity in politics while he veered toward involvement with the Communist Party which, in Britain and many other European countries, had become the sole political organization offering vocal and principled resistance against the rise of Nazi Germany. As was noted earlier, Haldane stressed that attempts at the eugenic removal of recessive alleles from a population would be ineffective because heterozygotes with one copy of that allele would appear normal. Mating between such heterozygotes would continue to produce homozygotes with the disease. Such matings became more likely with any increase in inbreeding.

Once gene sequencing became available, recessive alleles could be identified using DNA sequences rather than the appearance of individuals. The limitation that impressed Haldane disappears and transportation becomes less important. (Of course, this is a moot point. Independent of advances in genetics, human mobility has vastly increased since the 1930s.) From our perspective, what is most important is that the distinction between positive and negative eugenics maps naturally to our contemporary distinction between editing disease-causing genes and genetic enhancement. The next two chapters will discuss these two situations in turn. The salient difference between Haldane's era and ours is that we are not restricted to discouraging mating or aborting fetuses; CRISPR allows accurate editing of individual alleles to change them as we would want.

A WORKING DEFINITION

It is time to attempt a working definition of eugenics. This attempt will have a very limited ambition of capturing how the word is used today, especially by proponents of liberal or moderate eugenics (which differ on one critical issue, as we shall see below). We will not try to capture all past uses. We will view as *eugenic* any policy (or practice) that has the following three components. We will begin by assuming that we have targeted a human phenotypic trait that we wish to promote (that is, increase in frequency in the population) or discourage (decrease in frequency). Next, we will endorse conscious intervention in individual reproduction at the genetic level. Thus we implicitly endorse the assumption that, to the required extent, genes determine the presence of the trait. Finally (though this choice is less important than the other two parts of the definition), our purpose must be to change the future distribution of the trait in the population as a whole (and not be limited just to selected individuals).

Liberal eugenics adds a fourth component: achieving these traits must not involve any coercion. It must be left entirely to reproductive decisions made by parents. Society (presumably acting through the state) can have no role. *Moderate eugenics* does not accept this fourth component. While it also rejects coercion in general, it allows a role for the state to intervene in some eugenic decisions such as that of freeing a population from genetic diseases. According to moderate eugenicists, coercion is permissible in such situations to the same limited extent as in the case of enforcing mandatory vaccination. The discussion that follows will assume that what is at stake is liberal eugenics without necessarily rejecting the more expansive program of moderate eugenics. (If liberal eugenics cannot garner social acceptability as public policy, moderate eugenics is irrelevant.)

Thus, every component of a eugenic policy requires conscious choice. Every mating choice we humans make (short of cloning the entire population) results in affecting the future frequencies of traits in populations by creating new genotypes and changing the genetic profile of the population. But not every mating choice is eugenic according to our formulation because we may not be making conscious choices about genes and their future spread. This is in accord with our usual intuitions: we typically choose mates for reasons that do not concern the spread of genes over time. But consider the mating strategy of Fisher whom we keep on encountering. According to his daughter, thinking of the genetic profiles of potential mates formed part of his methodology for finding a suitable wife and subsequent reproductive choices. Fisher was happy to view his strategy as eugenic and it meets the requirements of our definition.

The requirements of our formulation are quite precise, perhaps too much so. Merely encouraging academically accomplished individuals to reproduce together, as was urged by the authorities in Singapore, does not fully satisfy our conditions. No definite trait was being targeted unless it is academic performance itself. But that does not seem to have been the Singapore government's intention; rather, academic qualification was supposed to be a surrogate for some other desired trait that remained nebulous. Most people will presumably accept that a policy of encouraging targeted marriages constitutes noncoercive intervention into reproduction at the genetic level. But that does not mean that a definite trait is being targeted. What matters in our context is that such a process seems far removed from targeted gene editing which provides the motivation for our formulation of eugenics. This is where this formulation diverges most from what eugenics was classically supposed to be, mainly about influencing patterns of reproduction.

Many societies have controlled reproductive patterns by elaborate rules, for instance, India's Hindus who required marriages to be restricted within the individual castes of their elaborate, rigid, and famously oppressive social system. Yet, even that system is not eugenics in our sense because no definite phenotypic trait is targeted and there is no conscious concern for the genetic profile of the population. (Hindu caste restrictions are also coercive in a way that would fall afoul of liberal or moderate eugenics if not all eugenics—but, as noted earlier, our discussion in what follows will generally be restricted to eugenics only in its liberal manifestation.) Of course, the ancient Hindus did not know anything about genetics when they set up their system of oppression thousands of years ago. But Galton was also hazy about heredity and his elaboration of eugenics only partly satisfies the formulation given earlier.

Not allowing coercion is central liberal eugenics. In emphasizing this, liberal eugenicists depart radically from most eugenic policies of the past, in particular those described in the first chapter. There is a well-known worry here: when does the role of a social institution itself become coercive? The coercion of the eugenic measures discussed in the first chapter is easy to discern and condemn: it was open and brutal. But there are much more subtle forms of coercion of which we must be wary. Economic difficulties may become coercive, for instance, when financial incentives are used to convince people to edit out "undesirable" genes. (It is within the realm of possibility that insurance providers in the United States will adopt such a measure.)

However, the problem of economic coercion is not unique to the context of eugenic or other reproductive decision in contemporary neoliberal societies. From the control of behaviors with adverse health effects such as smoking to the choice of unpleasant or socially stigmatized professions such as those in the sex industry, the potential for being coerced by financial need is omnipresent in our lives in capitalist societies. Finding solutions to this problem is beyond the scope of this book. It will require a much broader social discussion.

So, why would someone object to a noncoercive eugenics with obvious benefits to health and social well-being? Two factors *may* be operative (and the discussion that follows must be regarded as partly speculative). The first is a pervasive skepticism about social engineering in neoliberal societies. Eugenics has always had a social engineering component and continues to do so even in the formulation being used here: it includes a conscious goal to alter the future genetic profile of a society. Enthusiasm for eugenics among political liberals in Northern countries peaked in the 1920s, before the Great Depression, when the end of World War I (which was supposed to be "the war to end all wars") generated social confidence, and *confidence about science*, to an extent that is almost inconceivable today. That confidence disappeared with the failure of socialist planning in Europe and elsewhere. Today, the potential for eugenics leads to widespread social anxiety. That underlying anxiety may well explain why so many groups in the North reacted with such outrage when the Chinese experiments on CRISPR-mediated human embryonic germline editing were first reported in 2018.

The second factor adds to the intensity of this anxiety. Since at least the nineteenth century, there has been an implicit belief in most Northern societies that heredity is critical in determining human lives and futures. For the United States, this ideology is well documented. Many developments in modern biology contributed to this attitude. The original theory of evolution by natural selection—that of Darwin and Wallace—relied entirely on hereditary changes amplified by natural selection to explain the complexity and diversity of living organisms. Though Darwin subsequently flirted with the inheritance of acquired characters, ultimately, modern evolutionary theory rejected this type of inheritance and, also, any direct influence of environmental factors on inherited traits. The crucial figure in establishing that view, starting near the end of the nineteenth century, was a German zoologist, August Weismann, relatively unknown to the public today but one of the giants of the history of evolutionary biology. We have already been using one of the distinctions Weismann introduced, that between the germline and the somatic cells.

Weismann argued that not only is heredity restricted to the germline, but also that the germline alone carries specifications of what happens to the somatic cells, how—in our contemporary terminology—the genotype specifies the phenotype. Weismann attempted to support his theory by many experiments, including a notorious one in which he cut off the tails of mice over many generations only to find that mice continued to be born with tails. Weismann's views naturally dovetailed into Mendelian genetics after 1900 when Mendel's work began to be generally known. The hegemony of the gene consisted of the claims that not only are genes the stuff of inheritance but that they determine the most important features of humans (and other organisms). Moreover, they are passed on unchanged from one generation to the next endlessly except for very rare mutations which, in any case, would almost always hurt us.

The idea of the hegemony of genes and, therefore, of genetics over all other aspects of biology permeated Northern culture throughout the twentieth century. It even became a weapon deployed by the West during the Cold War after genetics was banned as bourgeois during the Stalin years in the Soviet Union. (This was the period of Lysenko's dominance over Soviet biology—recall our discussion in the last chapter of the fraud he perpetrated for several decades.) Indeed, the social power of genetics helped propel the sequence of events that led to the HGP even with its US three billion dollar cost. It is perhaps hardly surprising then that when social engineering appears to target the sanctity of the inherited genes themselves, the result is a deep anxiety about the future. Scientists, as Robert Cook-Deegan once put it, are playing God. Scientists would be interfering into what "God" had bestowed us. Implicitly, scientists would be claiming that they would be correcting God's work. No wonder many of us are uncomfortable.

INESCAPABLE EUGENICS?

Yet, since the earliest days of the HGP, many commentators have come to view eugenics at least as inevitable. In 1996, writing while the human genome was still being sequenced, the philosopher, Philip Kitcher argued that "[i]ntroducing molecular biology into prenatal testing will lead us to engage in *some* form of eugenics." (Kitcher's formulation was a little odd given that he was writing in the 1990s: quite some time earlier, molecular biology had already become part of prenatal testing for disease genes using DNA extracted from amniotic fluid. It is unclear why Kitcher thought of this only as a future possibility.) In any case, he was well aware both of the barbaric history of eugenics and of the morass we find ourselves in when we try to define it precisely. But he remained willing to use the term so long as we specify explicitly what we mean by eugenics. Eugenics, he felt, was "inescapable" even though it involved social engineering.

Kitcher proposed four criteria to evaluate different types of eugenics that could be envisioned:

> First eugenic engineers must select a group of people whose reproductive activities are to make the difference to future generations. Next, they have to determine whether these people will make their own reproductive decisions or whether they will be compelled to follow some centrally imposed policy. Third, they need to pick out certain characteristics whose frequency is to be increased or diminished. Finally, they must draw on some body of scientific information that is to be used in achieving their ends.

He concluded: "Practical eugenics is not a single thing. Human history already shows a variety of social actions involving four quite separate components, each of which demands a separate evaluation." For instance, eugenic policies that score most poorly with respect to the second, that is, those that embrace a lot of coercion may not be acceptable just for that reason no matter how well it satisfies the other criteria.

Kitcher's criteria differ from those incorporated in the definition we are using. What is now called liberal eugenics, he called *laissez-faire* eugenics. Kitcher was more willing to accept social intervention into individual decisions than most advocates of eugenics today, more so than even those who defend moderate eugenics and see some role for social enforcement of eugenic policies. Perhaps, the most striking difference between Kitcher's criteria and our definition is that he endorsed the idea that eugenics can be based on any scientific information rather than genetic information alone.

Suppose, for instance, social engineers allocate the number of children a family can have on the basis of economic status; the more money you have, the more children you are allowed to have. Their reasoning is that raising children requires money and the best organization of society would ensure that money is equally available for children when they are growing up. Let us suppose that this assumption is backed by credible economic data. This social policy may qualify as eugenic under Kitcher's criteria but would not do so under ours because our definition privileges genetics. In this, at least, our definition more closely matches what has traditionally been called eugenics. (Some versions of moderate eugenics also go beyond what is being defined as eugenics here since they allow for the manipulation of all aspects of heredity including epigenetic inherited factors.)

More to the point, even though Kitcher regarded eugenics as inevitable, he did not endorse any particular approach to it. In the chapters that follow, we will try to be much more specific. While arguing that eugenics must be inevitable Kitcher also made a very important point. Given what we know about genetics today, for instance, the rules of inheritance of genes and a fair amount of the connections between genes and trait, choosing not to use that knowledge during reproduction is itself a reproductive choice about the future distribution of genes in the human population. An example will make this clearer. Suppose that you know that you are carrying the allele for Huntington's disease. Also, suppose you are the genetic parent of an embryo. If you decide not to test it for the allele, you are making a conscious choice about the genetic future of the population. Implicitly you are deciding that there is no salient ethical difference between a population in which the allele exists and one in which it does not.

Or, suppose you test for the allele and find it present in the embryo. Then you decide not to use CRISPR-based (or some other form of) gene editing to correct the allele (assuming that safe genome editing has become possible). You may do so for a variety of reasons. You may object to "playing God." Or, given that Huntington's disease typically manifests itself well after the onset of adulthood, you may have decided that the value of the years without disease outweigh the cost of the years with it. No matter why you do it, you will have consciously affected the genetic future of the human population.

Let us make Kitcher's point a bit more stark: rejecting eugenic practices is itself a form of eugenics. As political activists have long pointed out in many contexts, choosing not to act is itself an action.

MATTERS OF ETHICS

The next two chapters will be concerned with the ethics of human germline editing using CRISPR. As we would expect, this potential use of CRISPR has been widely discussed by bioethicists. Much of this discussion has concerned gene editing in general and, though it now includes CRISPR, it had started well before the the advent of CRISPR because the prospect of gene editing has been around for a generation. CRISPR has introduced a new urgency to these discussions by making human germline editing an immediate live possibility. In the past, perhaps because the prospect of human germline editing seemed to be a problem for a distant future, many of those discussions remained mired in philosophical controversies that may only have marginal relevance to the practical questions we face today. Let us deal with and dispose these controversies here and try to glean from them anything that may be of value.

Bioethicists typically distinguish three broad theoretical frameworks for doing ethics. We will mention these briefly here, leaving a fuller discussion for the seventh chapter when more detail will be necessary in the context of a discussion of the ethics of genetic enhancement. Except in that case, a choice of a theoretical ethical framework typically makes no difference to how we think about the practical ethical issues that will arise in our discussions. (We will also gloss over many philosophical subtleties that similarly make no difference.)

The first framework takes a firm objective stand about right and wrong based on duties and obligations that individuals have to others, what they may or may not do to them. For instance, the seventeenth-century German philosopher Immanuel Kant insisted that we should not do to others what we would not have them do to us (and all others). This type of foundation for ethics is called *deontological*. (We will not encounter this word again; so, let us leave its origin aside.) A second framework, *virtue ethics* relies on developing good qualities in individuals, a tradition that goes back (in Western philosophy) to Aristotle in ancient Greece. The third framework, and the one most likely to be familiar to many of us, bases ethics on the consequences of actions or policies. *Consequentialism* argues for the maximization of good results and the minimization of bad ones. In its most basic form, it asks for the maximization of pleasure and the minimization of pain.

Following along the lines of reasoning that ultimately go back to the philosopher, Derek Parfit, some bioethicists have argued that the first two of these frameworks are irrelevant when we are confronted with eugenics based on gene editing. (What follows is *not* supposed to be Parfit's own view but draws on his reasoning.) Unlike the situation in which an individual is sterilized, or receives a reward for successful reproduction, no individual to whom we have obligations or whose development is concerned can be affected by germline editing: no such individual exists *at present* prior to that germline editing. The edited germline belongs to the future; the individual of which it may form a part is nothing more than a possible future individual who may or may not come into existence depending on whether an embryo comes to term.

Suppose an individual does come into existence, the preferences of such an individual may be affected by what we do now but we know nothing now of what that individual's desires, wishes, and so on, would be. The upshot of these observations is taken to be that we cannot say anything sensible about obligations toward such individuals or the development of individual qualities, and so on. The first two ethical frameworks are thus supposed to be irrelevant.

In contrast, consequentialism appears more promising. For instance, we can meaningfully talk about the increase or decrease of total pleasure or pain in some future situation even if we cannot otherwise meaningfully talk about the individuals involved. But consequentialism also has disturbing implications in the context of germline editing: we can imagine a technological situation in which we modify genes for receptors in the brain to enhance pleasure or decrease pain, for instance, in response to a genetic disease. For example, we could use genetic enhancement of pleasure to exceed whatever pleasure that could otherwise have been generated by removing the disease allele itself. Does it now become a more ethical choice to enhance such pleasure rather than address the cause of the disease? At present, this is an imaginary but not far-fetched scenario. We could clone people with genetic endowments that keep them always in a pleasurable state. Does that become a moral responsibility? (We will return to these possibilities in the seventh chapter.)

Where does this leave us? Does it mean that ethics, as practiced by professional philosophers, has little to contribute where it matters most, when we confront moral ambiguities about social problems we cannot avoid? Perhaps. Academic ethics may have made itself irrelevant to society by overspecialization and a focus on arcane practically irrelevant issues. The first observation to make is that the worries we just encountered are not unique to the question of CRISPR-mediated eugenics (or one using any other form of gene editing). With regard to the apparent irrelevance of the non-consequentialist ethical frameworks, the same issues arise whenever we talk about possible people, for instance, when we analyze our ethical responsibilities to future generations in the face of climate change or in any other context in which we have to worry about future generations.

From the Parfit perspective, we cannot ever know enough about the possible individuals who would comprise future generations to make decisions based on their preferences. It is the perspective that is problematic. It seems to me to be founded on patently false assumptions that Parfit's many followers simply have not questioned. Surely, we know enough to know that the next few generations will want clean air? Clean water? Ocean levels that are stable enough that coastal regions remain habitable? Precipitation patterns such that much of the world's agricultural lands remain in production? We know enough to know that we have an ethical obligation to combat climate change to the extent still possible and to prepare to adapt to the best of our abilities. Faced by the prospect of unbridled climate change, metaphysical worries about the status of future individuals seem beside the point.

Perhaps we need to assume no more than we just did, about the most obvious human needs such as pleasant and reasonably long lives with as little pain as possible, to argue that we have a moral obligation to free future human generations from genetic diseases to the extent that CRISPR (and our other technologies) now makes plausible. The next chapter will develop this line of reasoning but note its limitations in the face of developmental contingency and complexity. Our problem is not that we would have to edit the human germline; rather, it is that editing genes in the germline is likely to be a successful intervention only for a handful of genetic diseases.

The chapter after that will turn to genetic enhancement. Once again, an environmental analogy helps. Suppose we decide to plan the architecture of cities a few generations down the road. We assume that people will choose to drive (energy-efficient) cars, live in families, drive to work and shop, and by and large indulge in the passions of today's North American suburbia. Surely, this is silly, if not a recipe for a complete disaster? When it comes to choices of this type, we certainly do not know and cannot predict with any reasonable certainty what future generations will want.

In the North, nuclear families are a relatively recent phenomenon; most commentators trace their origins back to the Industrial Revolution and the social mobility it induced. However, nuclear families have largely been promoted as the "natural" and desirable unit of social organization only since the mid-twentieth century when the term was introduced by the anthropologist George Murdock who openly acknowledged the implicit atomic metaphor.

Today, many anthropologists and sociologists emphasize that nuclear families have *not* been the dominant human social organization through history, but that they may again give way to extended multigenerational families in the near future, and extended families may be a better form of social organization. Suburbia are a product of the 1950s and have led to myriad social and physical changes to the landscape, including the disappearance of biologically diverse habitats by spawning urban sprawl. It is hard to imagine that more environmentally aware and responsible generations will continue to organize their habitats as we have done in an age of waste and overconsumption. Even the automobile may lose the aura it has had since the 1950s.

These choices of earlier generations, particularly of the 1950s, were the result of the structure of power relations then, and our architectural choices today would continue to be similarly dominated by sociopolitical structural constraints even as we would try to be self-consciously critical of them. The same problem would manifest itself if we attempt the genetic enhancement of future populations. Our choices are more likely to reflect power relations today and the prejudices they generate rather than human universals about well-being beyond the negative eugenics of removing genetic disorders. At the very least, we should proceed with caution.

Chapter 6

The Elimination of Genetic Diseases

"This land is your land, this land is my land
From California to the New York Island
From the redwood forest, to the gulf stream waters
This land was made for you and me

As I was walking a ribbon of highway
I saw above me an endless skyway
I saw below me a golden valley
This land was made for you and me

. . .

As I was walkin'—I saw a sign there
And the sign said—no trespassin'
But on the other side . . . it didn't say nothin!
Now that side was made for you and me

—Woody Guthrie, "This Land is Your Land," 1940.

THE LONG REACH OF HUNTINGTON'S DISEASE

In 1962, when Bob Dylan released his first album, it included only two original songs. One of them, "Song to Woody," was a tribute to folk singer, Woody Guthrie, who is arguably the greatest songwriter the United States has ever produced. By then Guthrie, still barely fifty years old, had long been confined to a New Jersey institution suffering from Huntington's disease. With some effort, Dylan had been able to visit and sing for him. There was no hint of Guthrie's illness in Dylan's song; rather it was a celebration of Guthrie's wandering songwriting life over the years. Today, though few younger Americans seem to know who Woody Guthrie was, his songs live on. In particular, in school or church, most of this same younger generation have encountered "This Land Is Your Land," Guthrie's compelling 1940 response to Irving Berlin's cloyingly patriotic "God Bless America," even if today's most popular versions of Guthrie's song omit his more radical verses such as the last one quoted above.

In 1940, when Guthrie was only twenty-eight years old, his behavior and life had begun to unravel probably because of the onset of the early stages of the Huntington's disease that finally killed him twenty-seven years later. He had withdrawn into himself, become increasingly detached and cold toward his first wife, Mary, and their children, and had begun to exhibit occasional outbursts of unusual fits of temper. Woody Guthrie had inherited the Huntington's disease gene from his mother, Nora, who had died of the disease in 1929 at the age of forty-one. As a teenager, Guthrie had witnessed his mother's life slowly falling apart eventually to end with her institutionalization and subsequent death. She had lost control of her muscles at times and had suffered from deep depressions interspersed with outbursts of anger even before being sent to an institution. When Guthrie visited her for the last time in 1928, she did not even recognize her own son.

Fear of the disease seemed to have haunted Woody Guthrie ever since though his friends later recalled that he tried hard to live in denial. His career in music started in the 1930s in response to the Great Depression and the Dust Bowl that had driven him and many others from Oklahoma to California through Texas. He wrote songs about ordinary people and their struggles, about how migrants were unwelcome in California ("Do Re Mi"), and about outlaws who looked after the poor ("Pretty Boy Floyd"). During the time he spent in California, Guthrie worked the union halls and became a fellow traveler of the Communist Party though his enthusiasm for the Party flagged when Stalin and Hitler signed their short-lived but infamous nonaggression pact of 1940.

Subsequently, his career blossomed in New York in the early 1940s but he soon left the city to return to his wandering life all about the country. Even though signs of Huntington's disease were almost certainly already present, Guthrie continued to compose songs, record them, and perform extensively around the country through the 1940s. He remarried and his new work included a stunning collection of children's songs ("Songs to Grow On") inspired by his daughter Cathy Ann who was later killed in an electric fire at home in 1947. She was barely four years old. Woody Guthrie never quite recovered from his little daughter's death. By the early 1950s, Huntingtons had him fully in its grip and the last fifteen years of his life were largely spent confined to institutions, sometimes by his own choice.

One wonders how much more Woody Guthrie could have achieved had it not been for his illness. Many friends and relatives witnessed his intellectual and physical deterioration and the dissolution of his musical abilities; their pain is yet another aspect of the suffering brought about by Huntington's disease.

Today we continue to face the question of how much more future suffering there may come to be because of the continued presence of the Huntington's disease gene in the human population. It is a dominant gene; one copy suffices to cause the disease. But, today, we are also in a position to prevent that suffering: with CRISPR technology we finally may have the means to remove this gene permanently from the human population. So why not do it? That is the rationale for removing genes causing severe disease from the human population and it constitutes the most defensible form of eugenics. Once again, why not do it? Apparently we, as a society, remain unable to make that decision.

THE HE JIANKUI AFFAIR

Even though we may have a good rationale for removing disease genes from the human population, one thing is clear: society, and especially the biologists involved in CRISPR-based gene-editing research, are as yet far from ready to accept the conscious manipulation of the human germline. That much, at least, is clear from the furor caused in 2018 by He Jiankui who was at that time a biologist at the Southern University of Science and Technology in Shenzen (China). He had tried his hand at human germline editing. At the end of November, just before the beginning of the International Summit on Human Genome Editing in Hong Kong, He announced that he had used CRISPR-based techniques to successfully edit the germline in twin embryos in an attempt to confer resistance to HIV. The little girls were born in early November and, according to He, were healthy.

That this work emerged in China was not completely a surprise. Chinese biologists had been at the forefront of gene-editing research for several years. In March 2015, biologists from Guangzhou were the first to report editing genes in a human embryo, though they avoided controversy by using an embryo that was not viable. They were shortly followed in this type of editing by biologists from the United States and Britain. Because none of this earlier work involved viable embryos, it did not raise the hard questions and outrage that He's experiments provoked. In 2016, biologists from Sichaun University were the first to test the use of CRISPR-based techniques to edit genes in a human patient as a therapy for cancer. This was somatic cell gene editing and also raised no ethical hackles.

These developments motivated He to launch a project in June 2016 to edit the germline of human embryos with the intention of producing a live birth. In March 2017, he began recruiting couples with HIV-positive fathers for these experiments. One such couple led to the birth of the twin girls in November 2018. At the Hong Kong conference, He also announced the existence of yet another pregnancy with a germline-edited embryo which was supposed to come to term in August 2019.

He did not carry out his experiments in secret. He discussed his plans with biologists in the North including his former PhD adviser at Rice University, Michael Deem. No one appears to have tried to dissuade him very hard even though all of his interlocutors have since claimed that they tried to discourage him. But once He reported his results, immediately at the Hong Kong conference and elsewhere in the North, biologists reacted with fierce and vociferous condemnation of his experiments. Biologists claimed to be appalled at what He had done even though it was far from clear that he had violated any Chinese regulations or international standards. (Many biologists were also far from convinced of the scientific plausibility and value of He's experiment and we shall return to that issue later in this chapter.)

For instance, in response to He's announcement, the December 18 (2018) issue of *Science* contained an editorial by Victor J. Dzau, president of the US National Academy of Medicine, Marcia McNutt, president of the US National Academy of Sciences, and Chunli Bai, president of the Chinese Academy of Sciences that urged "international academies to quickly convene international experts and stakeholders to produce an expedited report that will inform the development of criteria and standards to which all genome editing in human embryos for reproductive purposes must conform, and to engage scientific bodies around the world in this effort." According to them, our ability to edit the human germline "has outpaced nascent efforts by the scientific and medical communities to confront the complex ethical and governance issues that they raise." The three Presidents were unequivocal in condemning He: "To maintain the public's trust that someday genome editing will be able to treat or prevent disease, the research community needs to take steps now to demonstrate that this new tool can be applied with competence, integrity, and benevolence. Unfortunately, it appears that the case presented in Hong Kong might have failed on all counts, risking human lives as well as rash or hasty political reaction."

This *Science* editorial is troubling at least on two fronts. First, it claimed that scientific developments have outpaced the formulation of policy. But, to the extent that this is true, it is a failure of the leaders of the genomics research community rather than any malfeasance on the part of individual researchers eager to explore the new possibilities opened up by CRISPR-based gene editing. As noted earlier in the book, the prospect of human germline editing has been recognized since the 1980s when it was actively discussed as a possible outcome of the Human Genome Project (HGP). Proponents of the HGP realized that the new genomics would have social impacts and the HGP included, right from the beginning, a program to study the ethical, legal, and social implications (ELSI) of the human genome. As much as 3 percent of the budget of the HGP was allocated to the ELSI program. Yet, almost thirty years later we find ourselves unprepared when presented with the immediate prospect of edited human germlines. Biologists could have done much more to insist that policy issues receive substantive treatment.

Second, the *Science* editorial implicitly acknowledged that there was at the time no pertinent rules in place at any level to regulate gene editing. (The state of affairs is not much better today.) This situation has the implication that He was being criticized for violating nonexistent rules which hardly seems fair. Moreover, in the *Science* editorial, the Presidents of the US National Academies of Science and Medicine offered to take the lead in developing regulations to cover human germline editing. However, in December 2015 these two academies had already hosted an international summit meeting on gene editing. One outcome of that meeting was a working group on the scientific, ethical, and social issues raised by new technologies, especially CRISPR. The group's report was published in 2017. Entitled *Human Genome Editing: Science, Ethics, and Governance*, that report did not call for an outright ban on human germline editing in embryos for research or otherwise. However, it recommended limiting both somatic cell and germline gene editing to prevent disease. This is what He is supposed to have done. What, then, are the international norms that He is supposed to have violated?

In the wake of He's experiments, some biologists went much further in their attempts to prevent human germline editing. Eighteen of them, led by Eric Lander of the Broad Institute and including almost all the luminaries of CRISPR-based gene-editing research, published a commentary in *Nature* arguing for an immediate moratorium on clinical human germline editing. Academy presidents got into the act again. Dzau and McNutt, this time joined by the president of the Royal Society of London, Venki Ramakrishnan, chimed in to endorse the biologists' call and proudly noted that their organizations "are leading an international commission to detail the scientific and ethical issues that must be considered, and to define specific criteria and standards for evaluating whether proposed clinical trials or applications that involve germline editing should be permitted." The Presidents were very happy that the World Health Organization was carrying out a parallel effort though one may wonder why such duplication was needed. Apparently, He's experiments had made the National Academies' 2017 report already obsolete because it was too permissive about germline editing. But is there good reasoning behind this change of position?

Chinese authorities were clearly embarrassed by the uproar caused by He's experiments. He was systematically investigated for compliance with all possible standards and regulations that could conceivably be relevant to his work. A news agency reported that the official conclusion was that He "intentionally dodged supervision, raised funds, and organized researchers on his own to carry out the human embryo gene editing intended for reproduction, which is explicitly banned by relevant regulations." No one publicly stated what these regulations were. In response to the investigation, He lost his job at the Southern University of Science and Technology. His career as a scientist was effectively over in China. But, to date, no official report has been released detailing what existing Chinese regulations He had violated in 2017 and 2018. In fact, China introduced draft regulations protecting embryos from gene editing only in May 2019 only after these developments had taken place.

By that time, He had disappeared from public view. At the end of 2019, Chinese authorities imprisoned him for three years and fined him the equivalent of US$ 430,000. Two of his collaborators were also convicted but received lesser sentences. But, once again, there was no public presentation of evidence of guilt, what *existing* regulations He had explicitly violated. There were rumors that Chinese authorities had knowingly funded his work. He may well have been a scapegoat sacrificed by Chinese authorities to curry favor with Northern institutions.

Meanwhile, in Russia in June 2019, molecular biologist, Denis Rebrikov, announced plans to target the same gene that He had focused on but to edit it differently in embryos which he would then implant in HIV-positive mothers to reduce the risk of transmission of HIV. So far, Rebrikov has not received the necessary permissions or publicly embarked on the project. It is unclear whether he will ever get the requisite permissions. Nevertheless, the genie is out of the bottle and will never be put back in again.

There is also no international (or scientific) consensus against human germline editing. For instance, twenty-nine European countries have signed and ratified the Oviedo Convention which bans germline editing but the list of signatories does not include Britain, Germany, Italy, or Russia. In the United States, the FDA requires an Investigational New Drug (IND) exemption for clinical trials involving transfer and gestation of a DNA-edited embryo. A US House of Representatives committee held a hearing in 2015 after which Congress passed an omnibus spending bill that explicitly prevented the FDA from using any of its resources to consider any IND application involving germline DNA modification. As a result, germline editing cannot proceed at present in the United States without it actually being illegal. Independently, the National Institutes of Health decided in 2015 not to fund any use of gene-editing technologies in human embryos thus effectively banning germline editing in practice.

Finally, in 2020, the US National Academy of Sciences, along with the US National Academy of Medicine and the Royal Society of London issued a report, "Heritable Human Genome Editing," concluding that existing technologies, including CRISPR, had not yet been demonstrated to be sufficiently safe for clinical use at this time. It limited its discussion to the use of human germline editing for disease prevention and did not broach the possibility of genetic enhancement which will occupy us in the next chapter. It recommended the initial restriction of the use of germline editing, after safety has been demonstrated, to cases in which a single dominant or recessive allele causes severe disease. It gave no reason to show that this was a wise choice perhaps believing it to be obvious. We will argue for a similar outcome below but only after providing a credible biological rationale.

THE ETHICS OF HUMAN GERMLINE INTERVENTION

It is time to move beyond gut-level protestations of horror and outrage and examine if there are genuinely ethical reasons to forgo germline editing in humans, especially for the purpose of eliminating genetic diseases. Why not proceed with editing the human germline to replace disease-causing genes with their more benign counterparts? Four "ethical" objections have been widely discussed and it behooves us to evaluate them.

Perhaps the strangest of these is the "playing God" objection which we have encountered before as an argument against sequencing the human genome. In the Judeo-Christian-Islamic tradition, we are supposed to be God's creation. It is supposed to follow from this premise that we have no right to alter ourselves, at least not in such a fundamental way as to change the future genetic profile of the human population. Though this objection seems to have impressed many people in the United States, particularly those who consider themselves religious, it is doubtful that it has any merit. For one thing, every time we choose to reproduce with someone, or choose not to do so, we affect the future genetic profile of the human population. This includes instances in which someone may choose not to have children because of a concern for passing down genes causing severe harm such as Huntington's disease.

Consider the case of Nancy Wexler, a geneticist who was instrumental in the identification of the Huntington's disease gene in 1983. She also happened to be the daughter, grand-daughter, and niece of people with the disease. Both Wexler and her sister decided not to have children for that reason. (This was at a time when the gene for the disease was yet to be identified; so, they did not have the option of being tested.) Were they playing God? And, if they were, was it wrong to play God in this way? That hardly seems plausible, Moreover, can a stricture not to play God ever override the responsibility to better the human condition by preventing the occurrence of such diseases? That seems absurd.

Not much better is the objection that our genome constitutes our common human heritage and any attempt to alter it shows culpable disrespect to that heritage. Just as we should not alter the Great Pyramid of Giza or the Great

Bath of Mohenjodaro, we should not alter the human genome. We could counter this argument by pushing this analogy further. A disease-causing gene is a change to the original DNA sequence similar to damage to a heritage structure. Just as we would almost certainly try to restore a damaged structure to its original form to the extent that we can (for instance, as we are now trying with the Notre-Dame of Paris), editing a genome to remove a disease-causing mutation would be restoring the original sequence. In either direction, the analogy is not very compelling. The human genome is a dynamic structure undergoing natural modifications all the time. Evolution occurs only because of this capacity for change. Even if we accept that our genome is our ultimate human heritage, that heritage is like a continually moving stream. Easing the flow of that stream by removing impediments hardly shows disrespect to that heritage.

However, the final two objections are more serious. When we alter the genome of an embryo that leads to a future child, we supposedly have altered that future child (and future generations) without obtaining any affected person's informed consent. The objection appears even more salient when we remember that a principle that constitutes one of the most important progressive ethical advances in medical research and practice in the twentieth century: participants in medical procedures (including experiments) must give informed consent to that procedure. This principle was one of the lessons we learned from the medical horrors perpetrated by German doctors from the Nazi era. But this objection has far too wide a scope to be definitive. Whenever we bring a child into the world, we do so without informed consent from the child. (When we attempt to conceive a child, that child does not even exist, let alone be in a position to give consent.)

Moreover, we recognize that a child even after birth (when its existence is beyond question) is still not in a cognitive position to give informed consent until close to adulthood. So, we typically allow parents to assume that role for children as we also do for fetuses and embryos. From this perspective, changing the genome appears very similar to other changes that parents may choose for in a fetus, for instance, to correct a congenital structural problem. Nevertheless, there appears to be an ethically relevant difference: changes to the genome are permanent and will persist into future generations in a way structural fixes won't. But, how real is this difference? After all, given the CRISPR-based technologies we would be using, future individuals may also be able to change the genome back to its original form. Most pertinently, given that editing would be preventing a severe genetic disease, it does not appear very plausible that future generations will want to revert to the original form.

The last point about CRISPR-based germline editing deserves emphasis. This type of human germline editing does *not* lead to a necessarily permanent change in the human gene pool because the edited germline can be changed back to the original form by further CRISPR-based editing. If editing is carried out using *current* CRISPR techniques, there is a complication. Each iteration of genome editing leaves some detritus in the genome in the form of DNA sequences disrupted by Cas9 (or some other DNA-cutting enzyme). But, because this detritus does not contribute to any phenotypic difference in the individual, we may choose to regard it as having no ethical salience. From a functional perspective, CRISPR-based gene editing is reversible.

The final objection is the most compelling. Trying to eliminate disease-causing genes presumes that these genes deserve to be eliminated because of how they affect the human phenotype. But, for the past three decades, disability advocates have been pointing out that those without disabilities should not be making a priori judgments about the quality of life experienced by those with disabilities. In many cases, those who are perceived as having disabilities view themselves only as different from other groups in society and not necessarily disadvantaged in such a way that they deserve to be altered from being who and what they are. Sometimes, those with disabilities choose to reproduce the difference and the cultures of life associated with them. Some deaf parents prefer to have deaf children. Hearing children of deaf parents often view deafness as a culturally different life style rather than as a disability. Thus, we should worry whether we are imposing our own values without due justification when we target a gene for elimination.

The past few decades have also been marked by social progress in the form of acceptance and accommodation of those with disabilities, for instance, in the United States with the passage of the Americans with Disabilities Act of 1990. Beneficiaries include those with genetic conditions. To some extent having many of these conditions have been destigmatized which is a welcome development. All this progress may well be lost if these attempts at cultural accommodation of human diversity are replaced with programs for gene elimination, especially through germline modification.

There is a two-fold response to these problems. The first is to be sensitive about the choice of language, for instance, possibly through the use of gene *modification* rather than *elimination* which, also, is more accurate in describing what CRISPR-based gene editing would do. Disability activists have a crucial role to play in establishing such linguistic conventions; even well-intended linguistic choices should not be made without their input. The time for beginning such a discussion has come. Should we even talk of *disability*? How do we present differences characteristic of a "disabled" condition in language that is not value laden in such a way as to encourage stigmatization?

Second, and more important, and here we will reluctantly continue to use the customary potentially problematic language until new conventions are proposed and accepted, the elimination of disease-causing genes should be restricted for the time being at most to a clearly demarcated small set of genes that are directly causally implicated in diseases for which there is no cure or adequate management. Huntington's disease and myotonic dystrophy are the kind of diseases that seem to provide good examples though the case of Huntington's disease presents complexities that we will shortly discuss. Moreover, these are diseases for which there seem to be no advocates suggesting that they are just differences that they would prefer to see persist in their children.

But there remains a residual problem that won't go away. Recall Woody Guthrie. There is no question that he would have much preferred a life without the prospect of Huntington's disease looming over him in early life and then, later in life, causing his mental and physical deterioration. But there are those who have maintained that the disease was also a source of his brilliance as a songwriter. According to one biographer, both his wife and one of his doctors speculated as much:

> And while it would be absurd to suggest that Huntington's disease *made* Woody Guthrie a brilliant songwriter, Dr. Whittier (and, later, Marjorie Guthrie herself) would wonder aloud if the disease hadn't worked like a drug on Woody, as a creative spur (in much the same that some artists use alcohol and other drugs), enhancing his natural rhyminess, forcing the brain to continually rewire itself as cells died, forcing new, wonderful, and unexpected synaptic pathways to open (which also led to some unexpected and not so wonderful behavior), forcing the brain to become—in effect—more creative to survive; and then, after a point, exhausted and starving for energy, the synapses and ganglia short-circuiting . . . preventing him from concentrating on anything, making him fidgety, antsy, causing him to lose perspective and, eventually, his creative sense of himself.

More recently, yet another physician, John Ringman, has developed the same argument even more forcefully. The claim seems to be that Huntington's disease contributed to Guthrie's brilliance.

So, perhaps, according to this line of reasoning, we should not eliminate its gene from the population. We justly revere people like Woody Guthrie and society would be better of with more of them. But does this speculative loss of brilliance outweigh the very real suffering that Guthrie endured (along with all those who also are victims of this disease)? And isn't it also the case that, in arguments of this sort, as with disability activists generally, the voices that matter are those of who are affected rather than those of others who choose to interpret their experiences for them? There is no doubt that Woody Guthrie would have preferred not to have had Huntington's disease. For me, that is the end of the story.

The conclusion to draw is that we don't have good ethical reasons to proscribe *all* interventions into the human genome. Moreover, we should also remember that we may even be blamed by future generations if we fail to act to eliminate from the population genes that cause severe disease just as we may be blamed for not eliminating pesticides and other environmental pollutants when we have the capacity to do so. Children may well blame parents for producing them with clearly defective genes which they could have removed shortly after fertilization.

THE LIMITS OF CURRENT SCIENCE

But this is where science intervenes to suggest caution, perhaps even skepticism. It reminds us how many of the promises of benefits of gene editing are no more than hype. Let us return to the claims of He Jiankui. He's results have never been published in a peer-reviewed journal (at the time of writing) and are impossible to assess satisfactorily. Nevertheless, He presented enough detail at the Hong Kong conference for scientists present at his talk to express strong doubts about the credibility of his results. Using CRISPR-based techniques, He targeted for the disruption of the gene for the surface protein, CCR5, of white blood cells. This protein is involved in the immune response. CCR5 was targeted because it is used by the HIV virus to establish an infection. He claimed to have disabled both copies of the *CCR5* gene which, because the allele shows dominance, is what would be necessary to confer immunity to HIV.

Though his slides were too dense with data to analyze properly during his hour-long talk, several scientists later claimed that his data showed that at least one of the girls continued to have one functioning *CCR5* gene which would mean that He's procedure had conferred no immunity to HIV. She remains as susceptible to AIDS as before. Moreover, He had not carried out any experiment to test whether cells from the girls were actually immune to HIV. Equally troubling, though He claimed that his use of CRISPR-based techniques had not generated any off-target mutations in the girls' genomes, he did not present even preliminary data to support that claim.

What all this means is that He presented no evidence to suggest that he had established the *accuracy* of his procedure: that it reliably targets the intended sequence correctly *and only that sequence*. The procedure would have been inadequate if only one copy of the *CCR5* genes was edited. Accuracy is thus a scientific criterion with normative consequences: if a procedure for gene editing is not accurate, then there is good reason to proscribe its use in human beings. CRISPR-based techniques are in general much more accurate for gene editing than the older methods they have replaced but they are still not perfect. Recent experiments have highlighted the severity of this problem. Three separate groups, two from the United States and one from Britain, have found that CRISPR can cause large DNA deletions or reshuffling on human chromosomes close to the gene targeted for germline editing. For every potential case of gene editing, particularly germline editing, which is much harder to reverse in an adult than somatic cell gene editing, accuracy must be established. This is what He failed to do and problems such as this are far more serious objections to his experiments than his failure to satisfy murky regulations about oversight.

He's experiments had even worse problems. He chose the *CCR5* gene for editing because some people carry a 32-base deletion known as a Δ32 mutation that deactivates the gene and provides an impediment to HIV infection. This mutation is well studied but the actual mutation that He introduced was different, and one for which safety had never been studied even in animals other than humans. Moreover, a recent analysis showed that people with two copies of the Δ32 mutation may have lower life expectancy than those without them. The mutation may also make people more susceptible to influenza and West Nile virus.

He thus appears not to have paid due attention to *safety*, perhaps the most important criterion that medical procedures must satisfy and one that is not restricted to gene editing. Further, even when a gene is clearly implicated in producing a phenotypic change, the question of safety remains paramount before editing should be contemplated because of the generally very complex relationships that exist between gene and trait. This is the problem of *specificity* which will be treated in some detail below because it has not received sufficient attention even though it is central to the question of when germline editing should be permitted. One-to-one correspondences between genes and phenotypes are very rare. Complexities abound and they lead to unintended consequences. There is *epistasis*, or interactions between multiple genes to produce traits, as well as *pleiotropy*, when a single gene affects multiple traits.

For instance, in humans, the *CCR5* gene interacts epistatically with genes for proteins called *β*-chemokines to mount an immune response to flaviviruses which include several tick-borne viruses as well as those that cause chikungunya, dengue, yellow fever, and Zika, besides the West Nile virus. He's intervention may have reduced the girls' resistance to all these diseases. Consider another example. The *SLC39A8* gene on chromosome 4 in humans is believed to play a causal role in producing both hypertension and Parkinson's disease. There is a common variant of *SLC39A8* that is known to decrease the risk for developing hypertension and Parkinson's disease and we may be inclined to consider gene editing to introduce this allele. Yet, a 2018 report points out that the same allele is believed to increase risk for schizophrenia, Crohn's disease, obesity and, possibly, other diseases. This is a textbook case of pleiotropy; whether *SLC39A8* also has epistatic interactions with other genes is not known but it would be very surprising if it didn't.

Even before the popularization of CRISPR-based tools as the preferred method for gene editing, some spectacular cases of pleiotropy were seen in livestock that had been subjected to germline editing. The gene *MSTN* codes for the protein myostatin that inhibits the growth of large muscles in mammals (including humans). Biologists at the Chinese Academy of Sciences eliminated the *MSTN* genes from cloned pigs in a successful effort to generate leaner meat. However, 20 percent of these gene-edited pigs had an extra vertebra. Biologists at Nanjing Agricultural University used CRISPR-based techniques to remove *MSTN* in rabbits to generate more meat. Once again, they were successful at that but fourteen of the thirty-four gene-edited rabbits were born with enlarged tongues. In another Chinese laboratory, when *MSTN* was removed from lambs, they had to use cesarean sections to birth them. In Xingiang, when CRISPR-based techniques were used to alter the *ASIP* gene in Merino sheep with the aim of creating breeds with specified wool color, the alteration decreased reproductive ability to such an extent that only a fourth of the implanted ewes carried offspring to term compared to normal ewes.

GENE SPECIFICITY AND LENIN'S BRAIN

The concept of gene specificity merits a discussion of its own. It was proposed in the 1920s but it never quite made it into the standard version of genetics even though we now recognize its importance. Specificity was introduced along with two other concepts, penetrance and expressivity, which are also highly relevant to how genes act and will be discussed below. The story of specificity involves two individuals, living and working in Berlin at the time, both of whom were colorful enough to have novels written about them. The story also involves Lenin's brain.

The first of these individuals was Oskar Vogt, a well-known German neuroanatomist who typically worked in collaboration with his French wife, Cécile Vogt-Mugnier. Since the 1890s, the Vogts' research program had been centered on the "architectonics" of the brain, that is, the size and form of the neural cells and their spatial organization, all of which, or so the Vogts believed, determined brain function. By the 1920s, when our story begins, Oskar Vogt was at the height of his profession and a director of the Kaiser Wilhelm Institute for Brain Research in Berlin as well as the editor of several prominent journals in his field. This prominence is what brought him into contact with Lenin's brain.

In the 1920s, in the immediate aftermath of World War I and the Russian revolution, the Western boycott of German scientists and the West's visceral hostility toward the Soviet Union had led to increased scientific contacts and collaborations between German and Soviet scientists. The Germans, including Vogt, frequently traveled to the Soviet Union. (Politically, Vogt was a socialist and these contacts with the Soviets were probably even more welcome for him for reasons that transcended science.) In the Soviet Union, by 1921, Vladimir Ilyich Lenin, the undisputed leader of the governing Communist Party, had begun to complain of fatigue, intense headaches, nausea, and insomnia.

By 1922, Lenin had placed himself under the care of German doctors. The same year Lenin suffered from two strokes and, in January 1923 Vogt, who was attending a neurology congress in Moscow, was called in as a consultant for Lenin's treatment. Despite the German physicians' best efforts, Lenin suffered a third stroke in March 1923 and subsequently died in January 1924. Vogt was asked to examine Lenin's brain which was sliced into sections for microscopic examination between 1925 and 1927. A special institute was set up for him in Moscow though he only supervised it from Berlin, meanwhile focusing his own investigation of Lenin's brain on a single section he was allowed to take out of the Soviet Union.

Vogt was aware that the Soviets hoped that his neuroanatomic evidence would show that Lenin had an outstanding intellect; in a preliminary report that he produced in 1927, he duly provided such an interpretation. However, in some articles and speeches, he also claimed to find similarities between Lenin's brain and those of criminals. These observations played into the hands of his many critics in Moscow who disapproved of having Lenin's brain analyzed by a foreign scientist and one who was beyond the Party's control. What Vogt did not know was the political significance of his report. In 1922, shortly after his second stroke, and presumably aware that his end may be near, Lenin had dictated a testament that was critical of several Communist Party leaders, especially of Josef Stalin, whom he wanted removed from the powerful position of general secretary of the Party.

After Lenin's death, what was politically at stake in the Soviet government was the state of Lenin's mind when he castigated Stalin. The Communist Party, which controlled all facets of the Soviet government, was embroiled in a factional dispute between Stalin and Leon Trotsky and their respective allies. While somewhat critical also of Trotsky, Lenin's testament nevertheless gave him glowing praise and there was no doubt whom Lenin would have preferred as his successor. Vogt's report suggested that Lenin was fully mentally capable at the time when he dictated the testament.

Though Lenin had intended his statement to be published, and it was circulated within the Central Committee of the Communist Party, it was not made public within the Soviet Union until 1956 (that is, after Stalin's death in 1953) though the *New York Times* published it as early as 1926. Stalin survived Lenin's indictment by suppressing the statement and eventually killing all his opponents. Trotsky went into exile and was eventually assassinated in México by one of Stalin's agents. Stalin went on to preside over one of the twentieth century's most brutal regimes. Meanwhile, in Berlin, the Nazis deposed Vogt from his academic positions because of his politics and he lost control over Lenin's brain much to the relief of Stalin and his allies.

Even if he was aware of it, and there is no evidence of that, Vogt was probably not much concerned with the political ramifications of his work. What is most pertinent to our story is that Vogt, though trained purely as a neuroanatomist, had extensive interests in evolution and genetics both of which he believed to be highly relevant to neuroanatomy. Between 1900 and 1920, as Mendelian genetics came of age, Vogt became convinced that psychiatric conditions were inherited and strongly influenced by genes. He hypothesized an analogy between the rules governing morphological variation in insects and the temporal sequence of the emergence of psychological abnormalities. Because he was dealing with psychological traits, Vogt was forced to accept ubiquitous plasticity, that is, variation in how a trait manifests itself, as opposed to the much more deterministic view of gene action endorsed by most Mendelian geneticists including those in Germany. At home, Vogt remained outside the mainstream genetics community.

In contrast, in the Soviet Union, probably because the official ideology of dialectical materialism rejected genetic determinism, a more nuanced view of gene action had become popular. Vogt found a receptive audience for his views at the prestigious Institute of Experimental Biology in Moscow. A distinguished genetics group had formed there; the two members of this group who most impressed Vogt were D. D. Romashov and Nikolai W. Timoféeff-Ressofsky. Both of them had discovered mutations in fruit flies that led to detectable changes in wing patterns but to different degrees. Timoféeff, in particular, had examined the effect of the same mutation in different genetically identical lineages or "pure lines" of flies. Only a fraction of the flies exhibited the effects of the mutation in each pure line but this fraction was characteristic of that line while it varied from one pure line to the next. In 1925, Vogt arranged for the publication of both results in a journal he edited, the *Journal für Psychologie und Neurologie*, which was an unusual venue for fruit fly genetics. Vogt also arranged for Timoféeff and his wife and collaborator, H. A. Timoféeff-Ressovsky, to move to the Kaiser Wilhelm Institute for Brain Research in Berlin so that they could collaborate with him.

In 1926, Vogt published a long paper interpreting the Soviet results. This paper introduced the three genetical concepts that concern us here: penetrance, expressivity, and specificity. First, he attributed a concept of "specificity" to Timoféeff even though the latter was yet to publish anything on the topic. In Vogt's words: "Timoféeff … correctly distinguishes different grades of specificity. The author then speaks of strong specificity when the mutation mainly always expresses itself through a strongly identical change of the same trait." He then went on to introduce a concept of "penetrance," the "tendency [of a mutation] to prevail." He showed, by example, that what he had in mind using the variation in the frequency with which a mutation appeared in Timoféeff's different pure lines. In other words, he viewed penetrance as the probability that a gene, if present, will manifest itself—this remains, essentially, our modern view of penetrance.

Finally, Vogt argued that a third concept had implicitly been used by Timoféeff. This concept was the "expressivity" of a mutation: this was the degree to which a gene manifested itself. It was distinguished from penetrance because a highly penetrant mutation could have low expressivity, that is, almost every organism with the mutant gene would show its effect but only weakly; and *vice versa* when very few carriers showed the effects of a gene but those that did showed it very strongly. What mattered to Vogt was that variability in specificity, penetrance, and expressivity would lead to a trait varying in intensity all over the spectrum of its possibilities even if it were the result of a gene mutation. Thus, with these conceptual innovations, Vogt was able to maintain that psychiatric conditions had genetic origins even when they were not inherited according to the rigid rules of Mendelian inheritance. Of course, as critics have pointed out, calling such traits "genetic" has no teeth—the relevant genes are only a small part of the causes of the origin of the trait. All that the new terms do is to enable an ideological insistence on the importance of genes in the face of recalcitrant data. But that is a story for another day.

It should come as no surprise that the Timoféefs endorsed Vogt's new concepts whole-heartedly. In an important paper, also from 1926, they devoted an extended footnote to the three concepts:

> The term "penetrance" (Vogt 1926) describes the ability of the factor [gene] of a phenotype to manifest itself. There are (strongly penetrant) factors, which always manifest themselves phenotypically, while others (weakly penetrant) come to be expressed only in a certain percentage of cases. The degree to which the factor manifests itself in the phenotype of the trait is described by the term "expressivity." The factors penetrance and expressivity can change independently of one another under the influence of certain other factors. Aside from dominance, penetrance, and expressivity the factors have a certain "specificity," i.e. the ability to always manifest themselves through a certain phenotypical trait. Various factors can have all just listed characteristics to various degrees.

There is some irony here. The Timoféeff's account of penetrance and expressivity is much clearer than that of Vogt who introduced those terms while Vogt gave a much clearer account of what Timoféeff seems to have meant by specificity.

Over the next two decades, penetrance and expressivity entered into the standard vocabulary of genetics, especially human genetics. Their adoption was helped by Timoféeff's rising reputation as a geneticist within the mainstream genetics community. In early discussions of the HGP, there was ample attention to the problems of incomplete penetrance and variable expressivity of genes that were supposed to be *for* a given trait. The problem was the following. The late 1980s and early 1990s were flushed with reports of discoveries of genes for complex human behavioral traits, for example, adolescent vocational interests, alcoholism, autism, bipolar affective disorder, male sexual orientation, neuroticism, obesity, reading disability, schizophrenia, spatial and verbal reasoning, and so on—Vogt would have been ecstatic.

These claims were reported with such great gusto in the popular press that we used to joke about a new principle of human genetics: "one adjective, one gene." What was ignored, especially in the popular press, is that each such scientific claim came with multiple caveats: that the gene was only one part of the causal story; that the identified gene had a very small overall effect on the trait; and that the gene typically had very low penetrance and weak expressivity. We then need to ask, if gene *a* is for trait *x*, but has low penetrance and weak expressivity, what is the sense in which gene *a* is *for* trait *x*? After all, most individuals with gene *a* would not show trait *x*—this is what low penetrance means, and even those who did show the trait would do so only to a very limited extent—this is what weak expressivity means.

This problem was never satisfactorily resolved. Genomics since the HGP has shown that, except in a vanishing few cases, for any complex trait, to try to say that any particular gene is *for* that trait, is misleading—a mug's game, to boot. But in the age of CRISPR, it is important to know, with high precision, that a targeted sequence is *for* the intended change before we modify the germline. Penetrance and expressivity are important, as we shall see, but what is critical is specificity as conceived by Timoféeff and Vogt. Gene editing of the germline requires that specificity be high. We turn to what that means.

A POLICY PROPOSAL

Unlike penetrance and expressivity, specificity never made it into the standard vocabulary of genetics. It did make it into one standard reference work, Rieger and Michaelis' first (German) edition of the *Glossary of Genetics and Cytogenetics*, and it was retained in the subsequent (mainly English) editions. However, this glossary defines specificity as the "quality" of a gene's action and goes on to claim that no clear distinction is possible between specificity and expressivity. No wonder that so little attention was paid to it. Though this glossary is generally reliable, the characterization of specificity it gave was faulty.

In the context of impending widespread germline editing in multiple species, it is time to revive the concept of gene specificity. The basic idea is that of Timoféeff as explained by Vogt: specificity measures the extent to which a gene does exactly what it is supposed to do, that is any change in the gene in any individual induces "a strongly identical change of the same trait." We only need to be clear about what constitutes high enough specificity to make germline editing permissible for disease genes. High specificity will subsume requirements about penetrance and expressivity. A proposed genetic change should be regarded as having the required high specificity provided it meets three conditions.

First, we should impose a *complete penetrance* condition: the induced genetic change must have complete (or almost complete) penetrance for the trait it is supposed to affect. This means that all (or almost all) persons with the edited gene will exhibit the changed trait. Recall that we decided to set the stakes high for when we would be willing to intervene in the germline. If this condition is accepted, we will not at present allow edits that affect just most individuals with the edited gene. We want to be sure that all (or almost all) individuals with the edited genes will show the trait change we are trying to achieve.

Second, we should impose a *constrained expressivity* condition: the induced genetic change must have an expressivity that falls within clearly understood boundaries of what is permissible. Simply being penetrant in almost all cases but only producing marginal improvements in a disease trait is not good enough reason for a germline intervention. For disease genes, the edit must produce a significant improvement from a diseased condition and, ideally, the complete absence of disease.

Third, we should impose an *exclusive effect* condition: the induced genetic change must only affect the intended trait and no other. This is the heart of specificity. It prevents the types of gene editing discussed earlier in which some fraction of individuals show unintended changes such as extra ribs and lolling tongues. Pleiotropy rules out specificity.

These three components of specificity are not independent of each other and must be approached simultaneously. An example will make this point. Watson and Venter, both of HGP fame, were the first two individuals to have had their entire DNA sequenced and to have made the results public in 2008 (though with some sensitive parts redacted). Watson's sequence revealed that he had mutations that lead to Cockayne and Usher syndromes. The mutations are recessive but Watson was homozygous for them: he had two copies of the mutant allele in each case. Both genes are known to have high penetrance. Watson should have both these diseases. Cockayne syndrome is rare and shows itself as microcephaly (abnormally small head size), accompanied by a failure to gain weight and grow properly resulting in a short stature and delayed developmental stages. Usher syndrome 1b (the version that Watson supposedly should have had) results in deafness and blindness.

Whatever we think of Watson, he doesn't have the symptoms of Cockayne and Usher syndromes. One possible way to make sense of this is to observe that the relevant genes had very weak expressivity in the context of Watson's developmental construction, his other genes and the history of his exposure to environmental determinants during early development. Thus, from the perspective of the human population, these genes, in spite of their high penetrance, did not have sufficiently constrained expressivity to show enough specificity to be acceptable targets for germline editing. Of course, the received view that these genes have high penetrance could also be incorrect. There are other possibilities. However, these issues are sorted out, when it comes to germline editing, we must ensure that we truly know that we are targeting genes with high specificity.

PROCEED WITH CAUTION

Our final proposal, then, is that we proceed with caution: in the context of recognized genetic diseases, human germline editing should be permitted for a set of genes provided that the protocol to be used satisfies the requirements of safety, accuracy, and specificity for the targeted gene. In general, specificity will be achieved in the case of those traits, including disease traits, that are controlled by a single or very few loci so as to exclude epistasis as an important factor. Pleiotropy is inimical to specificity because it violates the requirement that changes in the gene should only affect the intended trait and no others. *CCR5*, *MSTN*, and *ASIP* are examples of genes that do not satisfy the specificity requirement. This proposal would not have permitted He's experiments.

The urge to edit the human germline has largely come from our long experience with diseases caused by single genes, that is, malfunctioning alleles at a single locus, particularly when they are dominant. By and large, these genes and their diseases will satisfy the specificity condition. Myotonic dystrophy remains a good example. Problems of penetrance complicate Huntington's disease in an unusual way. The mutant "gene" for this disease is not a single allele: there are hundreds of variants all implicated in generating the symptoms associated with it; dynamic changes in the genome during reproduction generate that variety. It will be instructive to look at this process in more detail.

Huntington's disease is caused by expanded and unstable repeats of a cytosine-adenosine-guanine (CAG) triplet in the gene that normally codes for a protein that has been named "huntingtin." (The exact function of this protein remains unknown though it is believed to be tied to long-term memory.) The CAG triplet specifies the amino acid residue glutamine in the protein it encodes. The more repeats there are, the longer the chain of glutamine there will be in the mutant huntingtin molecule. Normal individuals, that is, those without Huntington's disease have between 11 and 26 CAG repeats. Those who have more than 40 repeats of this triplet are supposed to have a 100 percent probability of developing the disease. We then have complete penetrance. There have been cases with as many as 250 repeats. A number of repeats between 36 and 39 is associated with variable penetrance (from 25% at 36 repeats to 90% at 39 repeats). The clinical course of the disease varies widely even for those with the same number of repeats.

Symptoms of Huntington's disease typically develop between thirty-five and forty-five years of age though the age range of when the first symptoms appear varies from two to eighty years. For instance, Woody Guthrie had two children from his first marriage who developed juvenile Huntington's disease (that is, showed symptoms before the age of twenty) though both had lived to be forty-one years by the time of their deaths. In general, the number of CAG repeats is believed to be predictive of the age of onset of the disease with higher numbers leading to earlier onsets.

Does Huntington's disease satisfy our specificity criterion? Suppose we accept a penetrance of 90 percent as a reasonable threshold for intervention through germline editing. Then only those mutant disease alleles with thirty-nine or more repeats would be potential targets for germline editing. But what about those who do develop the disease with fewer repeats but, nevertheless, have severe symptoms? Shouldn't their suffering have been prevented? Moreover, part of the dynamics of Huntington's disease over generations involves an increase in the number of repeats during reproduction, particularly during meiosis and the formation of sperm in males. So, not intervening even in situations where a variant allele at present contains only a small number or repeats over twenty-six will lead to the continued persistence of severe Huntington's disease in the population in the future because of the amplification of these repeats each generation.

It thus appears that our proposal is too restrictive. However, we may want to accept that limitation for the time being because, as we embark on human germline editing, which is unprecedented (with He's attempts being an exception), it seems wise to restrict ourselves to cases in which there is no plausible doubt about what, precisely, a germline intervention would achieve and why it seems appropriate—indeed, desirable. We may then gradually expand our repertoire of permissible germline edits in the future after we become better aware of consequences including, especially, unintended consequences if any, of the results of editing. We should also note one sense in which our proposal is quite permissive in some ways. We are not restricting editing to only those diseases for which there is no other option, that is, where intervention at the genetic level is the last resort. We are also not requiring high expressivity for the disease gene itself in the sense that the symptoms of the disease must be severe. We are only constraining the expressivity of the *edited* gene to be such that it results in an indubitably healthy phenotype. Of course, in practice, it is unlikely that any one would propose even somatic, let alone germline, gene editing unless a disease has severe effects.

LIBERAL AND MODERATE EUGENICS

So far, in this chapter, we have not considered who would make the decisions to edit a germline with the intention of removing a faulty gene from the population. Most of this book concerns only liberal eugenics and, from that perspective, germline editing would be part of individual reproductive decisions, for instance, when parents decide to choose the genetic profile of an embryo that would eventually be brought to term. Such a process directly influences the genetic status of an individual. It will not remove a gene from the entire population if there remain parents who choose not to eliminate a gene. Even if such parents can eventually be convinced to eliminate the gene, its elimination from the population as a whole would still be a slow process.

So be it. Proceeding slowly, and with caution may be a virtue when we are embarking on policies for which we only have very scary precedents, that is, eugenics as practiced in the past. However, liberal eugenics is not the only eugenic option in town. We can imagine elimination of disease genes emerging as matter of social policy. Moderate eugenics embraces such a policy. For instance, if we eventually achieve socialized medicine, social cost considerations may suggest that certain genetic diseases be prevented through germline modification rather than be managed through expensive treatments. However, there may be very good reason to resist any such move on the ground that reproductive decisions should remain a matter of personal choice not to be trumped by social policy. What we most need is a public debate on these issues.

Chapter 7

Designer Baby Delusions

J. B. S. Haldane (while speaking about eugenics to a journalist): "Crew, what is the perfect man?"
F. A. E. Crew (who happened to be walking by): "There isn't any. Define us a heaven and we'll tell you what an angel is."

—William E. Laurence, 1932, "Not a 'perfect man' in Haldane's Utopia."

WHY NOT GENETIC ENHANCEMENT?

Why stop with the elimination of undesirable genes from individuals and populations? Why not move on to the genetic enhancement of desirable qualities? Much of the debate over gene editing, whether it be in somatic cells or the germline, has assumed that a move to enhancement crosses some vivid ethical line of acceptability. But does it? And, if so, does ethics suggest we ban genetic enhancement or *encourage* it?

Is genetic enhancement akin to giving a proper education to our children? Or is it more like pumping them with steroids during puberty or even earlier with the hope of making them better athletes? For that matter, is there anything ethically unacceptable about pumping children with steroids so long it has no deleterious side effects? Why is consuming steroids worse than the special diets coaches enforce on high school football players in Texas (and elsewhere) to develop their bodies into the right size and shape? Why is genetic intervention worse than enforcing diets or even injecting steroids? (*If it is*, as most commentators seem to assume without apparently feeling any need to offer arguments for that assumption.) Or is enhancement only problematic if it interferes with the germline? We will have to navigate through a minefield of thorny issues.

THE NORMAL AND THE ENHANCED

The first problem we encounter is that the concept of enhancement turns out to be much more complicated than initially expected when examined in detail. Enhancement, in a biomedical context, is supposed to go beyond restoring or maintaining capacities a person would ordinarily have. In other words, whereas the customary goal of medicine is to restore these ordinary capacities (or prevent their attrition in the case of preventive medicine), enhancement is supposed to do something more. Thus, a philosophical encyclopedia article defines enhancement as "biomedical interventions that are used to improve human form or functioning beyond what is necessary to restore or sustain health." The 2003 US President's Council on Bioethics similarly identified enhancement with biomedical interventions that went "beyond therapy." A minor problem with this definition is that it presumes "beyond" to consist only of going over the limits of normal functioning only in the right direction. For instance, suppose we encounter people who have skins that are excessively sensitive to sunlight. So, we intervene medically to reduce their sensitivity. But we intervene to such a degree that they no longer have any sensitivity to sunlight at all along with all the benefits it brings such as the capacity to synthesize vitamin D. This would hardly be called enhancement. We would have begun by restoring normal functioning by decreasing sensitivity but then would have gone beyond the normal limits in the wrong way.

However, the real trouble with the definition is that what constitutes *health* is itself open to dispute. At one end of the spectrum are those who hold that health can be given a completely scientific definition based on the statistical distribution of measurements for different bodily parameters in all functional members of our species. From this point of view, what we call health does not reflect our values at all; it is as objective a feature of our bodies as our height or weight. For instance, if we were interested in hearing ability, we would assess the hearing performance of a representative sample of all normal members of our species to establish what healthy hearing requires. Unhealthy status would be attributed to those falling outside this range in the wrong direction.

But what is normal hearing? Does it include perfect pitch? If it does, then changing someone's hearing to produce perfect pitch would not constitute enhancement. If it does not, doing exactly the same thing would become enhancement. But whether or not perfect pitch is within the normal range is not a scientific issue. It requires a value judgment on our part just as, going back to one of our discussions in the last chapter, whether or not hereditary deafness is a disability (rather than a difference in condition) is also a normative judgment, that is, a question about our values rather than about facts in the world. The "objective science" definition of health is dishonest: it claims without proof that there is no value judgment required when determining health. It simply hides the fact that values form part of this judgment by referring to normal functioning without spelling out what *normal* functioning means.

The boundary between attempts to enhance human capacities or just to restore normal functioning is also blurry in another way. Consider adult ADHD, which is characterized by symptoms such as being easily distracted or restless, having difficulty with making plans, or being impulsive and chaotic in behavior. Most researchers believe that about 1–3 percent of adults experience ADHD but estimates range from 0.05 percent to 7 percent. Clearly, it is difficult to draw a clear boundary between ADHD and normal restlessness, impulsivity, or lack of concentration. Many of those who are diagnosed with ADHD do not view it as a disorder, let alone a more serious disease. For them, it is part of acceptable normal variation in human behavior. If that is correct, treating ADHD using methylphenidate or other psychostimulants, as is commonly done, could be considered enhancement because it goes beyond restoring or maintaining normalcy. When we encounter examples such as this, enhancement seems to be ubiquitous. We should be encouraging enhancement in many situations.

Though there have been the usual claims of genes "associated" with ADHD—and the meaning of such claims will be dissected below—no one has seriously proposed a genetic solution for the purported problem. Even if there are genes that play an important role, there would be too many of them, each with a tiny influence on ADHD, and possibly strong influences on other traits. There would be too little genetic specificity to make genetic intervention plausible. Indeed, there would likely be so many genes involved that genetic intervention at the present state of knowledge, even with CRISPR technology, would be absurd.

Let us turn to a different example which most people take to have a clear genetic basis: skin color. We expect genetic influence because it is clearly inherited though we should not expect a single gene to be involved because of blending: children tend to have a skin color intermediate to those of the parents. (We also do not see families that suggest a single gene is involved. For instance, when the two parents have different skin color, we do not see a quarter of their children exactly resembling one parent while the remaining three-fourths resemble the other.)

We also have compelling evidence that discrimination on the basis of skin color is widespread across the world and is an important component of racism. A 2011 study in the United States found that light-skinned African-American women received shorter prison sentences than those with darker skins. A 2015 study found that white interviewers regarded light-skinned African- and Hispanic-American interviewees as more intelligent than their darker-skinned counterparts even when there was no other basis for such a judgment.

One study found a significant gap between career opportunities for light- and dark-skinned women in India. Yet, another study found that, in India, dark-skinned Indian women had virtually no success at online dating. So, it should come as no surprise that there is a widespread desire to get lighter skins.

The preferred method of getting lighter skins in much of Africa and Asia is through the use of skin-lightening creams. Even though many of these creams are known to be dangerous to health, a 2018 *Guardian* article estimated that the global skin-lightening industry was worth US$ 4.8 billion in 2017 and was expected to grow to US$ 8.9 billion by 2027. The World Health Organization estimated that 40 percent of women in China use skin-lightening creams; in India that number increases to 61 percent; and in Nigeria to 77 percent. While political liberals from societies that claim to disavow racism might find the practice distasteful, those who use skin-lightening creams often view the practice as no different from using cosmetics or other standard methods to change their physical appearance. For users of skin-lightening creams, the benefits seemed to outweigh whatever harm that may come from skin lightening through the reinforcement of societal "color prejudice," to use an archaic term for racism.

The use of skin-lightening creams is probably the most widespread biological enhancement technique the world has ever seen. Given the extent of discrimination that occurs on the basis of skin color, condemning the practice of skin lightening seems downright irresponsible unless it is accompanied with tangible and credible measures to end discrimination on the basis of skin color. One can even make an argument that skin lightening, so long as it is carried out safely, should be encouraged to counter discrimination so long as it continues to exist. In India, where arranged marriages are still the norm in most regions and take into account the physical appearance of potential brides, parents may well feel a moral obligation to lighten the skins of their daughters to secure better marriages for them. Again, the benefits seem to outweigh the cost of reinforcement of prejudice.

There is no question that skin lightening meets our definition of enhancement. Indeed, if we suggest that it consists of restoring or sustaining normal health, then we would credibly be open to charges of blatant racism. We will turn later to the genetics of skin color. Right now, we have a more pressing problem. If skin lightening makes us ethically nervous, shouldn't that worry carry over to other to other traits that are endorsed by proponents of biological enhancement? Or, is the case of skin lightening really different from, say, increasing intelligence? If so, why?

Take another example, relevant to both China and India, the potential to alter the sex of an embryo and, in particular, to turn it male. In India, abortion of female fetuses is perceived to be a social problem to such an extent that fetal sex determination is illegal. But, we can point out that the existence of severe discrimination against women in almost all facets of contemporary Indian society is good enough reason to choose not to bring another little Indian girl into that world. Going further, one could potentially even argue that it would be a case of wrongful birth. Wouldn't would-be parents even have an obligation to alter the sex of the embryo to its advantage, that this would be a highly desirable enhancement? We can no longer avoid two questions: What traits? Who decides?

LIBERAL EUGENICS AND GENETIC ENHANCEMENT

Proponents of liberal eugenics have been vocal in their enthusiasm for genetic enhancement for over two decades, dating well back into the period before the emergence of CRISPR technology supposedly made the prospect tangible. Two figures stand out among the cadre of liberal eugenicists, John Harris of the University of Manchester and Julian Savulescu of the University of Oxford. Harris has argued that a commitment to genetic enhancement logically follows from our willingness to use other technologies to promote our welfare and capacities: that willingness is supposed to make us morally obligated to pursue genetic enhancement. Perhaps this is the reason why many Indian parents feel they should lighten their daughters' skins. Savulescu has pursued a very similar line of reasoning in even greater detail and has been prolific in his publications. As the director of the Oxford Uehiro Centre for Practical Ethics, he has also publicly promoted genetic enhancement to a wide audience.

What motivates both Harris and Savulescu is the promotion of well-being. According to the latter, if we have an obligation to treat and prevent disease, including genetic disease, we also have an obligation to similarly enhance traits that would promote well-being. "Enhancement," Savulescu has argued in an article promoting better breeds of people, "is a misnomer. It suggests luxury. But enhancement is no luxury. In so far as it promotes well-being, it is the very essence of what is necessary for a good human life." Moreover, Savulescu insists, enhancement can and

Table 7.1 Savulescu's Criteria for Ethical Enhancement

What is an ethical enhancement?
It is in the person's interests
It is reasonably safe
It increases the opportunity to have the best life
It promotes or does not unreasonably restrict the range of possible lives open to that person
It does not harm others directly through excessive costs of making it freely available (but balance against the costs of prohibition)
It does not confer an unfair advantage
It does not place that individual at an unfair competitive advantage with respect to others, e.g., mind reading
It does not reinforce or increase unjust inequality and discrimination—economic inequality, racism (but balance the costs of social/environmental manipulations against biological manipulations)

What is an ethical enhancement for a child?
All the above plus:
the intervention cannot be delayed until the child can make its own decision
the intervention is plausibly in the child's interests
the child consents if competent

must be ethical. Table 7.1 lists his criteria for ethical enhancement. All enhancement attempts must meet the first eight conditions listed there; those carried out in children have to meet three more (which are relatively uncontroversial). We will soon look at that table carefully. Returning to Savulescu's arguments, after promoting these criteria for ethical genetic assessment, he adds a ringing exhortation:

> Our future is in our hands now, whether we like it or not. But by not allowing enhancement and control over the genetic nature of our offspring, we consign a person to the natural lottery, and now, by having the power to do otherwise, to fail to do otherwise is to be responsible for the results of the natural lottery. We must make a choice: the natural lottery or rational choice. Where an enhancement is plausibly good for an individual, we should let that individual decide. And in the case of the next generation, we should let parents decide. To fail to allow them to make these choices is to consign the next generation to the ball and chain of our squeamishness and irrationality.

We will further explore the rationality of human genetic enhancement in the next section.

The argument from well-being is not the only one that has been offered in defense of genetic enhancement. The late John Robertson of the University of Texas staunchly defended procreative liberty, "the freedom to decide whether or not to have offspring and to control the use of one's reproductive capacity." For Robertson, procreative liberty established a presumption in favor of free choice in reproduction that could, however, be overridden by more compelling ethical reasons when necessary. Liberal eugenicists can make this position even stronger by going beyond the mere existence of a presumption to make the freedom of parental choice more absolute so as to deny society any control over reproductive decisions. To get to genetic enhancement, we must add to Robertson's list the freedom to choose freely what kind of children to have. If procreative liberty is absolute, this would include the freedom to have children with genetic disabilities, for instance, the gene for Huntington's disease. Such a choice would be hard to defend from criticism. But, if (with Robertson), we only assume a presumption for free choice, we could accept that the harm that would result from this choice gives ample ethical reason to override the freedom to choose what kind of children to have. Enhancement would in this way remain permissible but the pursuit of genetic disease would not.

Nicholas Agar of the Victoria University of Wellington in New Zealand offers yet another libertarian argument in defense of genetic enhancement: the freedom of parents to influence the direction of their children's lives. Agar's reasoning is more sophisticated than what we have so far encountered. Unlike most of those who have written on this topic, he does not fall afoul of an illegitimate genetic reductionism (which we will discuss in some detail below), that is, the assumption that genes alone can account for the presence of a trait in organisms. Rather, he accepts a developmentalist perspective that we have the traits we have because of complex interactions between environmental and genetic factors, both of which are, in principle, equally important. Now, we take it for granted that parents not only should have the freedom to manipulate environmental factors to enhance their children's lives, but that they have an ethical responsibility to do so. Parents should feed their children as well as they can. They should educate them as well as they can. They should facilitate any apparent gifts, whether it be in music, mathematics, or soccer. But, if environmental factors are on par with genetic ones, shouldn't parents also enhance the genetic factors as well? If it is permissible, indeed desirable, for us to try to enhance a future child's intelligence by manipulating a mother's diet during pregnancy, shouldn't we also try to provide the embryo with the best available genes for intelligence? Agar's argument seems to be persuasive.

RATIONALITY OF GENETIC ENHANCEMENT

To examine the rationality of genetic enhancement, we must take a close look at why it is supposed to be desirable, that is, the ethical assumptions on which enhancement is supposed to be grounded and judge whether these assumptions are warranted. Harris and Savulescu rely on consequentialism, that an action or policy should be judged on the basis of what it produces as outcomes. For them, the good to be maximized is well-being. They say very little about what constitutes well-being though their writings and other pronouncements suggest that it must be a subjective feeling connected to happiness, what Robert Sparrow has called "a warm inner glow, as it were." But if that is what we should do, why should we bother with the traits that liberal eugenicists typically promote, for instance, cognitive ability or preferred physical appearance?

As Sparrow has pointed out, we should directly engineer people with brains designed with appropriate neurotransmitters generating an overwhelming sense of well-being, people who "go through life suffused in a warm bath of serotonin, dopamine, and opioids." Alternatively, if we are so able, we could manipulate genes so as to ensure that negative feelings do not arise, for instance, by mutating putative genes associated with depression. Genetic enhancement would thus no longer be dependent on the traits that liberal eugenicists had originally promoted. Rather, genetic enhancement would take a short cut to the genes most directly promoting well-being. Indeed, why should parents not edit as many plausible genes as they can to achieve that goal even if it means that the child would end up having very little genetic similarity with either parent? (Of course, it may be the case that the parents' well-being may then take a severe hit because they will know that they lack a genetic connection to the child but we could do a cost–benefit analysis to preclude genetic changes beyond that tipping point.)

The only way to avoid such a fate would be to provide a convincing account of well-being that goes beyond this subjective feeling of contentedness. But would such an account support genetic enhancement? There is reason to be skeptical. Suppose, for instance, that we fall back upon traditional philosophical theories of good character development. (We are now in the realm of virtue ethics, which is an alternative to consequentialism as we saw in the last chapter.) Let us assume that working hard toward some goal is one aspect of desirable character. We then stipulate that well-being includes a requirement that persons show concerted effort to achieve goals that are at the edge of their abilities. It follows that we should not genetically enhance these abilities, or at least not perfect them. Rather we should encourage children to make concerted efforts to achieve the desired ends. Of course, liberal eugenicists could deny such an account of well-being. But, then, to avoid Sparrow's objections, they have to provide a different convincing account of well-being. So far they have not done so, relying on intuitive assumptions of what many people find desirable, for instance, that improved intelligence is desirable.

But once we fall back on such intuitions, liberal eugenicists are in trouble. As our earlier discussion shows, whether it be racism or sexism, genetic enhancement by maximizing well-being seems to entail that we succumb to every social prejudice generated to any extent by the presence of genes. Pandering to prejudice is a sure way of increasing well-being: in any society a majority of individuals are likely to feel better when their social prejudices continue to be affirmed. Of course, Savulescu will point out that his ethical conditions in Table 7.1 explicitly rules out "reinforcement or increase of unjust inequality and discrimination."

The trouble is that, if we take this requirement seriously, it is difficult to find traits promoted by liberal eugenicists that should still be enhanced. Take height: routinely enhancing height would likely lead to increased discrimination against short people. Take weight: routinely enhancing appropriate weight relative to height would very likely lead to discrimination against fat or very thin people. Indeed, the same problem would presumably afflict every physical trait. Well-being in many societies would require genetic manipulation, *if possible* (and the evidence is not compelling), against homosexual orientation simply because of discrimination against those with that sexual orientation. Yet, trying to alter sexual orientation through germline editing of an embryo would reinforce exactly that discrimination. Suvalescu's *ethical* enhancement seems in practice impossible.

That still leaves us with the problem of complex mental and behavioral traits such as intelligence. But now we are back to our old problem of the connection between these traits and well-being. In the case of intelligence, it is very likely that improving some cognitive faculties when they are much less than typical would lead to better lives. But that is not enhancement; it does not transform the cognitive ability beyond the normal level. But, once we propose to enhance beyond the normal functional level, there is no evidence, nor any other compelling reason to believe, that we would be increasing well-being of that individual. It is possible, though, that a society composed of many people with enhanced intelligence would be better off than one with very few such people. However, if we argue for enhancement on the ground that society *as a whole*, rather than the individual, would benefit from it, we are no longer in the realm of *liberal* eugenics with its emphasis on individual freedom. Rather, as Sparrow has pointed out, we are back to the eugenics of the past that led to involuntary sterilization and mass murder.

Savulescu's criteria for ethical enhancement ask for trade-offs to be computed between genetic approaches to changing socially valued traits to social and environmental manipulations. As Sparrow and many others have pointed out, for most such traits, whether it be increasing cognitive performance or decreasing racial or gender-based prejudice, in spite of all our biological advances, social interventions remain the most effective means of increasing well-being. The nest way to enhance a child's academic performance is to provide good nutrition, a good home environment, and competent schooling. Indeed, if well-being is what is at stake, liberal eugenicists have so far given very little reason to pursue genetic enhancement at least at present or in the foreseeable future. What they have inadvertently accomplished is provide compelling reasons for an enhanced pursuit of social justice.

Robertson's and Agar's more libertarian arguments do not fare much better. Neither of these arguments claim an absolute freedom of choice about what kinds of children parents should be permitted to have. So, we are again left to pick and choose which traits should be enhanced. As in the case of Harris and Savulescu, we are faced with the problem that the traits most likely to be targeted for enhancement are those that would pander to the social prejudices of the day. If we admit that pandering to prejudice is ethically impermissible, for Robertson and Agar that provides reason enough to override libertarian claims of parental freedom to enhance traits in their children. For physical traits, the problems with pursuing enhancement remain those that we discussed earlier whether the trait be height, weight, skin color, or other aspects of physical appearance. For behavioral traits, once again, we are only left with liberal eugenicists' intuitions that traits such as intelligence should be enhanced. But, as we shall shortly see, intuitions are not enough.

Suppose that liberal eugenicists insist that their widely shared intuitions are sufficient to serve as guides for policy. Then they run afoul of competing contradictory intuitions that strongly reject *genetic* enhancement of *all* traits. Recall that Agar argued for parity between environmental and genetic manipulation for the enhancement of traits. If a pregnant woman can go on a special diet to enhance the intelligence of a fetus, this is supposed to be no different than her choosing to edit the genes of an embryo using CRISPR-based technology after *in vitro* fertilization and before implantation. The trouble for Agar is that there is a widespread intuition—at least as widespread as the one that promotes enhanced intelligence—that germline editing in this case should remain impermissible even though the special diet is not. We are not claiming that Agar is relying on such intuitions in any fundamental way. We are only pointing out that appeals to intuitions would not give him the results he needs in favor of genetic enhancement.

The source of the intuition that environmental interventions are almost always permissible but genetic interventions are not (except in the case of debilitating disease) is the perceived permanence and irreversibility of germline interventions. Environmental manipulations result in *acquired* changes and, as we have noted earlier, a core assumption of twentieth-century biology has been that such changes cannot be inherited; in contrast, genetic changes can be transmitted across generations indefinitely into the future. This assumption has been challenged by the emergence of epigenetics since 2000 but the extent of this challenge is unclear. Work in epigenetics has shown that some acquired features can be transmitted from parent to offspring at least for a few generations. Typically, these involve changes in molecular structures on chromosomes attached to the DNA but not changes in the DNA sequence itself.

In any case, the emergence of epigenetics has not yet had much influence on discussions of human germline editing; the epigenetic results so far obtained have not been enough to challenge the strong intuitive presumption that the human germline should not be manipulated. Proponents and critics of eugenics agree that choosing to edit germlines changes the course of evolution. We saw in the last chapter that this potential has horrified some critics to the point of challenging germline editing even in the case of genes responsible for debilitating diseases.

This is where CRISPR potentially comes to the rescue of liberal eugenicists. Because CRISPR-based gene editing is so easy, any change that is introduced can also be reversed (as we saw in the last chapter). There is nothing necessarily irreversible about it. If we make a mistake when we edit the germline to enhance a trait, we can edit it back to the original state. In practice, this may not be easy: we would have to trace every person into whom the edited genes have been transmitted. But, in principle, it could be done. How seriously we should worry about tracking down these individuals is a question of how optimistic we are about our technology to track recipient of CRISPR-edited genes. Liberal eugenicists are technological optimists *par excellence* and would refuse to let this worry change their course. But technological prowess will not make the problem of pandering to social prejudice go away.

Suppose we introduce some genetic change today, say, an enhanced ability to play chess because that is what is supposed to indicate appropriately enhanced intelligence. Two generations later, people generally realize that this was a mistake; it only reflected the prejudices of certain societies in the early twenty-first century. So, we (that is, our descendants) reverse the change with a lot of difficulty. (It turns out that finding every person who had inherited those genes is hard, and convincing each of them to acquiesce to that change is harder because they have all become addicted to chess.) But, for all our troubles, how can we know for sure that the new decision against chess is not itself a product of the prejudices of its own time and community? After all, playing chess better is not like living a life free from myotonic dystrophy.

PLANNED HUMAN OBSOLSESCENCE

Genetically enhanced persons would, to some extent, be products of technology, and the extent to which they are technological would depend on the extent of the editing, that is, on the number of introduced changes and the phenotypic severity of each such change. For many commentators, every time technology is used to modify a human being, that person becomes closer to an artifact. Some of these commentators have fully embraced this development and have even glorified the advent of a "trasnshuman" future. For others, especially those of a conservative bent, the idea is anathema and to be denounced in the strongest possible terms. Leaving aside what we think of transhumanism—and, except in works of (intentional or unintentional) science fiction, it is not imminent in spite of CRISPR and other new technologies—Sparrow has recently raised another disturbing possible consequence of genetic enhancement.

Imagine parents decide to enhance a child's intelligence (or any other desirable trait) with the best available alleles at the early embryonic stage (when CRISPR-based intervention would be most appropriate). Suppose that the procedure is a complete success and the child is growing up with a sense of pride in the accomplishment. Ten years later, the same parents decide to have another child, once again with genetically enhanced intelligence. But, now better alleles are available, that is, those that do more to enhance intelligence with higher specificity. Naturally, the parents choose to use them.

Sparrow argues that the older child, as an artifact, will now have become obsolete and, more importantly, *feel* obsolete. How would this affect a person? How will it affect society? Will we end up in a situation when each genetically enhanced child, growing up, knows that obsolescence is part of each gene-edited child's inevitable future? What would this do to our concept of personhood? The logic of enhancement that leads to obsolescence in this way also subsumes human beings under a technological dynamic in which technological developments determine what a human being should aspire to be as a person. What does this do to human nature?

Though Sparrow has brought these issues into sharp focus in the context of CRISPR-based technology that has made human germline editing plausible, he is not alone and also not the first to express such worries. Almost twenty years ago, the philosopher Jürgen Habermas raised similar concerns. Habermas argued that genetic enhancement blurred the distinction between "the born" and "the made"; human beings would become products rather than persons. Embryos would become systems to be manipulated rather than nourished and this would transform what constitutes personhood. These developments would be drastically detrimental to human freedom because individuals would no longer understand themselves as being the authors of their own choices.

WHERE IS THE SCIENCE? THE CASE OF INTELLIGENCE

So far, we have spent a lot of time on philosophical arguments for and against human enhancement, in particular, genetic enhancement potentially enabled by CRISPR technology. But before we proceed any further, it is important to pause and ask what biology has to say on the matter: is genetic enhancement plausible? Possible? Or have we been indulging in flights of fancy? Let us start with a trait very dear to liberal eugenicists: intelligence. According to current biology, or even the biology of the foreseeable future, can we plausibly genetically enhance human intelligence?

It will serve us well to begin with the work of Robert Plomin, a celebrity-academic psychologist in the United Kingdom who is one of the most vocal purveyors of the idea that genes are a major determinant of intelligence and, thereby, of the rest of intellectual and professional performance. He is also willing to wade into social controversy. Plomin is a regular performer on radio and TV talk shows in Britain. In 2015, in one of these shows, he was asked whether there was any link between race and intelligence. "In general I've felt softly, softly is a better way to go," he replied, sidestepping the question. Plomin was the primary source of a full-blown genetic determinism about intelligence that animated a leaked 2013 essay of over two hundred pages prepared by Dominic Cummings, a close adviser of Michael Gove, the British Secretary of State for Education (from the Conservative Party). According to Cummings: "Work by one of the pioneers of behavioral genetics, Robert Plomin, has shown that most of the variation in performance of children in English schools is accounted for by *within school* factors (not *between* school factors), of which the largest factor is genes." Those who endorsed Plomin's views included Boris Johnson, then mayor of London, who has since gone on to greater things.

One of Plomin's modest societal proposals was the creation of a "genetically sensitive school" designed to match children with their presumed genetic capacities. Writing with Kathryn Asbury, Plomin promoted this vision in a 2013 book, *G is for Genes: The Impact of Genetics on Education and Achievement* :

> We aim to treat all children with equal respect and provide them with equal opportunities, but we do not believe that all our pupils are the same. Children come in all shapes and sizes, with all sorts of talents and personalities. It's time to use the lessons of behavioral genetics to create a school system that celebrates and encourages this wonderful diversity.
>
> One way of helping each and every child to fulfill their academic potential is to harness the lessons of genetic research . . . It's time for educationalists and

> policy makers to sit down with geneticists to apply these findings to educational practice. It will make for better schools, thriving children, and, in the long run, a more fulfilled and effective population. That's what we want schools and education to achieve, isn't it?

Asbury and Plomin cannot be blamed for lack of ambition. Their vision for the school is truly breathtaking:

> The site we choose for our genetically sensitive school will be enormous, more like a small university campus than a traditional school. It will have to be this size to hold all of the facilities it needs to accommodate and all of the options it needs to provide. It will serve the community around it, and we will make it so appealing and so successful, and we will foster such a pleasant environment and such a wonderful reputation, that every child of every faith, every race, and every social background will want to be educated there.

It is hard not to wish them well.

Asbury and Plomin do not explicitly make a case for genetic enhancement. They claim the high ground by supposedly caring for and catering to all children, whatever their genetic endowments. But their discussion also makes it clear that genetic differences cause differences in achievement for a wide variety of skills, especially those like intelligence that carry social acclaim with them. Intentionally or not, they whet the appetite for liberal eugenics, that is, parental manipulation of genes to generate brighter children.

Though Plomin was trained as a psychologist, his adopted area of expertise in behavioral genetics (which has not been uncommon within parts of psychology since the mid-twentieth century). So, we turn to that discipline to assess the plausibility of his claims. As its name suggests, behavioral genetics studies the role of genes in generating typically complex behavioral traits in a variety of species though the only one of interest to us here is *Homo sapiens*. But, to start with, how does behavioral genetics choose what is a *trait*? For instance, is the number of teardrops shed in a lifetime a trait? If not, why not?

What is a trait? Unfortunately, this question does not have a technical answer within genetics, that is, there is no theory that determines what is a trait independent of what biologists choose to study for any reason. For instance, biologists study features of organisms that appear to be functionally important (such as body weight, height, and a large variety of morphological as well as physiological features). They also often study those features that appear striking, especially if they are also passed on from generation to generation (for instance, the number of fingers in a human hand or tumbling behavior in pigeons). Typically, they choose features with sufficient stability and regularity so that they can carry out repeatable experiments with them which is probably why the number of tears shed in a lifetime is not of much interest.

Plomin and those of his ilk simply assume that intelligence is a trait and that it is appropriately measured using IQ tests. Liberal eugenicists do not question this assumption. Yet, other psychologists such as Ken Richardson have pointed to the remarkable malleability of what intelligence is supposed to mean. Social psychologists, Gabriel Mugny and Felice Carugati, carried out an experiment in which a large number of claims about intelligence were provided to parents. They were asked to assess, with a score between one and seven, the extent to which each claim was true about intelligence. Many of these claims contradicted each other but many parents were content to affirm contradictory claims about intelligence. For instance, there was equal agreement on the following two claims: that the "development of intelligence is the gradual learning of the rules of social life"; and that the "development of intelligence proceeds according to a biological program fixed at birth." Another psychologist, Robert Sternberg, along with several collaborators, questioned a large number of psychologists what they took intelligence to be. Of twenty-five possible attributes of intelligence, only three were mentioned by more than a fourth of these psychologists.

In fact, some languages, such as those of Sanskritic origin, do not even have a word that captures the connotations for "intelligence" in contemporary Northern societies; Bengali, for instance, has a word for "wise" (*buddhiman*) and one for "clever" (or "sharp"; *chalaak*) but not for "intelligent." Psychologists have also pointed out the ideological role that intelligence plays in contemporary Northern societies. Mugny and Carugati note its cultural provenance: "Intelligence, if such a thing exists, is the historical creation of a particular culture, analogous to the notion of childhood." Sternberg is even more damning: "intelligence is invented . . . it is not any one thing . . . Rather it is a complex mixture of ingredients. The invention is a societal one." This means that there is ample reason to reject the view that there is a single biologically well-defined trait that should be called intelligence.

Plomin and like-minded researchers claim to measure intelligence through IQ tests. Now, it is logically possible that, even if intelligence as a whole should be viewed as a set of many different capacities, IQ scores capture one of these accurately (and IQ thus becomes a concept of some scientific value). So, we turn to the question of what IQ is. Both critics and proponents of IQ are legion and some critiques such as Stephen Jay Gould's *Mismeasure of Man* have achieved iconic status in the philosophy of biology. Critical race theorists such as David Gillborn have persuasively argued that IQ has always been a tool of racial discrimination and subordination, no matter whether the targeted "races" were southern Europeans, Jews, Africans, or all people of color—in fact, the target has predictably varied with the dominant

Northern politics of the era. For instance, in the early 1920s, IQ testing of would-be immigrants at Ellis Island in New York City determined that 83 percent of Hungarians, 79 percent of Italians, and 87 percent of Russians were feebleminded.

It is probably no accident that Plomin went "soft" about IQ and race in 2015. Two decades earlier, when incendiary racial claims about race and IQ were promoted in a book, *The Bell Curve*, and had become a matter of public debate, Plomin was one of those who signed a high-profile piece in *The Wall Street Journal* that claimed to represent mainstream psychologists' agreement with the empirical claims in the book. These claims included not only a difference between IQ scores for whites and blacks but also claims that they showed no sign of converging and that they were largely independent of economic class. Almost no data supported these grandiose claims. In response, a large number of genuinely mainstream psychologists also went on record to reject all of these claims.

Let us leave these controversies aside for the time being and, instead, focus on how IQ tests are designed. The idea that there is a single general internal power or capacity for mental ability goes back to Galton in the nineteenth century. However, Galton's subsequent influence on the IQ story is negligible. That story really begins around the turn of the twentieth century with Alfred Binet in Paris who designed what has come to be regarded as the first IQ test to measure intellectual performance of children. Unlike those who followed in his footsteps, Binet was not concerned with theorizing about the capacities he measured. Rather, his goal was to identify children who required remedial help in school irrespective of the causes. Binet wanted to optimize the provision of this help. Things became strikingly different when IQ testing was exported to the United States, particularly in the hands of Terman whose goal was to identify "mental defectives" who should be removed from society.

What matters most in our context is how Binet designed his test. As Richardson has noted, for Binet, intellectual performance involved at least general knowledge, memory, imagination, attention, comprehension of sentences and synonyms, aesthetic judgments, and moral judgments, a list far more comprehensive than the ones used by those who have followed him to the present time. To test for these capacities, Binet devised a vast array of questions that he proceeded to administer to children. Next, and this is the critical move, he selected a subset of these questions to be part of his final test using two criteria. The first was whether the average performance in answering a question became better with age; if so, it was supposed to give some indication of a child's intelligence. The second criterion is the one that has characterized IQ tests ever since: Binet asked whether children's performance on a question matched their teachers' judgment of their intelligence.

Now, psychologists take it for granted that teachers are very good at predicting students' academic performance. Given how Binet selected his questions to match teachers' assessments of students, it follows that IQ can also correctly predict academic performance. Newer IQ tests, especially those developed in the United States, followed Binet's methodology and are typically referred to as Stanford-Binet tests. The methodology has been manipulated by test designers in a variety of interesting ways. For instance, in 1937, it was discovered that girls on the average scored a few points lower than boys in the Stanford-Binet tests that were in vogue at the time. Test designers debated whether to allow this difference to persist and ultimately decided against it. The questions generating this difference were duly removed and we had a gender-neutral test *by social construction*. Other manipulations are equally informative. Tests always include a large number of questions that most people get right and very few that most get wrong or very few get right. The result: a bell curve for IQ—there is no more to that celebrated shape than an artifact of test construction.

So, where does this leave us? We are supposed to be measuring an invisible power, general intelligence, that some researchers such as Charles Spearman have called *g*. What we get are visible data from IQ tests that consist of a bunch of scores. These visible scores are supposed to measure the invisible power. This situation is routine in science. Think of the invisible power called gravitation. The visible data are the positions of the sun and planets at different times. Using these data, we can make many inferences about the power of gravitation, how strong it is, how it changes over large distances, and so on. But we can only make these inferences because we have a superb theory that connects the invisible to the visible. This is Newton's theory of gravitation, one of the most successful scientific theories ever formulated (though it has had to be somewhat corrected by Einstein's work). But what about *g* and IQ scores? We have no theory at all. IQ scores are well behaved only to the extent that we have introduced our intuitions about what they should do by manipulating the construction of the tests to serve our purposes.

It should come as no surprise to us that, beyond academic performance, most analyses show that IQ scores are poor predictors of job performance. They are not even good predictors of cognitive ability, for instance, the mental capacities used by regular bettors at a racetrack. Then there is the "Flynn effect": mean IQ scores have risen around the world, year by year, decade by decade. In some countries, it has risen by fifteen points over three decades. All that is likely to be going on is increasing similarity between what test questions ask and what people are regularly doing more and more throughout the world because of increasing globalization. People are probably not simply getting smarter. At least not to this extent.

GENOMICS AND IQ

Does this mean that liberal eugenics is delusional if it expects to enhance intelligence through genetic manipulation? Well, Plomin and others in his camp still have one card left to play: even though many aspects of IQ may be artifactual for the reasons we have just seen, there may still remain an underlying core, *g*, that is stable and respectable because it has a biological, indeed, a genetic basis. In fact, the claim that IQ is genetic has been around for more than a century though its popularity has waxed and waned. The claim has always been controversial and especially so when it is coupled with a claim of presumed biological differences in IQ of different races, a claim that has no scientific basis whatsoever.

We will not elaborate on the politically charged question of race here while acknowledging that there is good reason to view IQ testing as having been driven by racism throughout its hundred-year history. We will see that, even independent of all racial considerations, the genetics of IQ turns out to be illusory. So we do not even have to address issues that put together IQ, genetics, and race together in an unholy trinity. Historically, throughout the twentieth century, attributing IQ to genes became less popular when attempts to find the implicated "candidate" genes failed spectacularly in spite of dedicated efforts; the claims became more popular when new technologies were invented to detect genes with supposedly more subtle effects on traits. Following this pattern, the launching of the HGP resulted in renewed claims that IQ is genetic. In recent years, the pendulum has swung the other way.

Returning to Plomin, he places his own work as promoting what he calls the "new genetics of intelligence." These new "findings" rely on the latest technology for supposedly finding genes related to IQ, the genome-wide association studies (GWAS) that we already discussed in the third chapter. Though most early attempts to use these studies to tie IQ to genes were unmitigated failures, according to Plomin and Sophie von Stumm, recent results are supposed to have turned the situation around. What is at stake in these disputes is the concept of *heritability*, more specifically, *broad heritability*, *H*, of a trait. IQ is supposed to be genetic because it has a high heritability (of around 50 percent). The trouble is that the concept of heritability, and what (if anything) it shows about genetics, has been contested by geneticists ever since it was introduced in the 1940s. Critics, and there are legions of them within the genetics community, hold that future generations will have as much truck with heritability as ours does with phrenology.

What heritability means will be central to our discussion of genetic enhancement of intelligence. But, first, a point of clarification: though what follows will be quite critical of the concept of heritability, we are not suggesting that genes have nothing to do with cognitive capacities. Rather, these capacities, like all other aspects of our biology, depend critically on our genes just as they do on the environment of development, that is, the history of interactions between genomes, cells, and higher-level factors internal and external to the body. In this mitigated sense, we are products of our genes but we are equally products of our environment. More accurately, we are products of our history, both individually and societally. That history includes both biology and culture.

Now, returning to heritability, let us start with some well-defined trait, unlike IQ, that varies continuously across a population. This means that the trait can take any possible value between its observed minimum and observed maximum; it is not discrete in the sense of being restricted to a small set of exact values. Height and weight are good examples; blood group is not because it can only take one of a handful of values. We will use height as an example.

We can measure the height of each member of our population. From these numbers, we can calculate the average, or *mean*, height as well as its *standard deviation* which shows how much variability it has in our population. The square of the standard deviation is known as the *variance* (of height in the population); it, rather than the standard deviation, is the quantity most often used in science because it has some nice mathematical properties. This variance is the *phenotypic variance*, V_P, of the trait (in our case, the height). It is a quantity we can directly measure.

To get to heritability, we must ask: what fraction of the phenotypic variability is due to genotypic variability in the population (that is, variation between individual genotypes)? Since this variability can also be captured by a variance, and symbolized V_G (broad) heritability can be defined as the ratio $H = V_G/V_P$, a number that can vary between zero and one. But, how is V_G to be measured? If we can manipulate experimental populations, we can at least approximately estimate V_G directly. For instance, suppose a population of plants is grown in such a way that all individuals experience the same environment. There are important subtleties here: we cannot grow all the plants in exactly the *same* environment because they cannot all be grown on the same spot. Rather, we use everything we know about plant growth (ambient light and temperature, soil composition, acidity, humidity, *etc.*) to make sure that each plant experiences the same value for all relevant environmental variables. (Already, there is an important disanalogy with IQ: even after all these decades of research the exact environmental factors influencing IQ remain unknown.)

After controlling for environmental factors in this way, if we measure the variance in plant height in various fixed environments, we have an estimate of V_G for our experimental population and can calculate H with reasonable confidence. The trouble is that H, so estimated, depends critically on the genotypes in the population and the environmental values that the plants were grown in. If the genotypic composition of a population changes, as it would in the next generation unless every member is cloned, H would change; it would also change if the environment varies beyond what it was in the last generation (in our example, if it changes in any way).

If these aspects of H already suggest that its utility is very limited, there is more to come. Consider the human population in a landmine-infested region, for instance, parts of Cambodia in the 1990s. Now, presumably, we would all agree that all human beings have two legs for genetic reasons. Given that mutations that change the number of legs are very rare, for our hypothetical Cambodian population, we would reasonably assume that $V_G \sim 0$. Yet, there would be many people with only one leg in this population, because they would have had the misfortune of stepping on undetonated mines. This means that $V_P > 0$. We thus get $H = V_G/V_P = 0$ for a trait (having two legs) that should clearly be regarded as genetic.

Now, consider an example that goes the opposite way. Consider a population, and these are not hard to find in the United States, in which the only language spoken is English. Let us assume that there is small fraction of this population that does not speak English because of genetic cognitive impairment. This means that the only variability we have in the population is genetic. So we would have $V_P = V_G$ and $H = 1$ even though the trait "speaks English" is not encoded in genes. Heritability is a strange measure indeed.

What matters most in our context is that what heritability at best measures is the fraction of the variability in a population that can be attributed to genotypic factors for exactly that set of genotypes and the range of environments to which the population has experienced. It says nothing about the mean value of the trait in the population, let alone the value it has in an individual. Thus, high heritability by itself says nothing about the genetics of a trait (as the earlier examples were intended to show). Most importantly, high heritability by itself provides no guidance on whether a trait is malleable through environmental manipulation.

Consider height in humans. It is known to have a very high heritability; according to several analysis, it is around 80 percent. Going by how the heritability of IQ has been used by the proponents of a genetic basis for intelligence, the high heritability of height should be taken as evidence that the mean height of human populations cannot be changed except through genetic enhancement. Yet, the mean height of human populations throughout most areas of the world has been increasing each generation. It has increased not due to genetic changes but because of better nutrition. Heritability is a statistical measure but it is not one that points toward a causal story.

Estimating heritability in human populations is non-trivial because ethics prevents us from setting up experimental populations. So, indirect methods are used and these are fraught with problems that delight critics and are typically ignored by heritability enthusiasts. One of the best-known of these methods is twin studies. A popular strategy is to study identical twins reared apart. For any trait when a pair of these twins shows differences, those differences can be attributed to the environment. Let us call the variance V_E. Now, if we assume that $V_G = V_P - V_E$, then we can estimate $H = V_G/V_P$ from studying these twins. Analyses of this sort have typically returned $H = 0.5$ (or 50 percent) for IQ. The trouble is that this calculation assumes that there is no interaction between the genes and the environment.

But identical twins look alike and, in most cases, that leads to other people treating them similarly. Children treated in the same way in widely different environments will be psychologically more similar than those treated differently. Thus, for cognitive traits, it is implausible that there ever is no interaction. There are other problems too: no analysis has ever convincingly shown that twins reared apart experience very different environments. Typically, adoption agencies try to match many aspects of the backgrounds. How important this is we don't know: we have a very poor understanding of what environmental features are relevant for IQ.

This is where GWAS is supposed to have changed the game. It apparently provides an entirely new way of estimating heritability from the strength of the association of IQ with regions of the genome. Plomin and von Stumm review the evidence and claim that each new study since 2015 has given a higher value of V_G. But the best value they could report in 2018 is a heritability of 10 percent which is well below the 50 percent estimate from the earlier problematic results. For this 10 percent result, they report IQ to be associated with about a thousand different loci or regions in the genome. Even if we accept these results—and they remain controversial—this does not give much teeth to the clam that IQ is genetic. If thousands of loci can be involved, and each has a very tiny effect, it is hard to imagine a trait that would not be genetic. Moreover, the problems with interpreting heritability remain, most importantly, how it provides no guide to the value of a trait has for an individual (or even for the mean value in a population).

Plomin is, of course, aware of the subtleties with heritability and why it cannot say anything about an individual person's capacities. His response is to appeal to another counting technique that has emerged in the wake of GWAS: what he calls *genome-wide polygenic scores* (GPSs). Unlike heritabilities, Plomin and von Stumm claim, "GPSs predict intelligence for each individual." How is this supposed to work? Suppose there are a number of different loci that are correlated with IQ scores and we have the numerical values for these correlations. At each locus, we have a number of alleles that increase IQ. The standard way to calculate a GPS score is to multiply this number of alleles with the correlation value for that locus and add it all up for all the loci that are involved. The trouble is that to view this number as the causal genetic contribution to intelligence requires a very vivid imagination.

For one thing, we are adding up weighted correlations but have no basis for regarding the correlations as relative causal contributions. For another, we are adding up alleged contributions of alleles and loci as if they never interact with each other. Genetically, this is fantasy. If alleles never interact with each other we have no dominance and, whenever we look at well-studied genes, some amount of dominance is all over the place. Moreover, we are concerned with hundreds of loci that are identified by GWAS as being associated with IQ. Each of these loci is supposed to have a very small effect on (a low correlation value for) IQ and also affect scores of other traits. Yet, for all the complexity, it is supposed to be the case that the different loci associated with IQ have no influence on each other. Plomin and von Stumm seem to acknowledge some of these problems when they also urge caution in interpreting GPSs. But this note of caution gets lost in their bloated rhetoric.

As Noah Rosenberg along with several collaborators have pointed out, polygenic scores come in two stripes polygenic risk scores and genomic polygenic scores as exemplified by Plomin's GPS. For both types, the computation begins with a GWAS. More specifically, as they explain:

> Over the past 15 years, genomic analyses have identified thousands of genetic variants that contribute statistically to variation in complex phenotypes, traits that have complex patterns of inheritance and that are affected by large numbers of genes in combination with environmental factors . . . In a typical genomic study of a complex human phenotype—a genome-wide association study (GWAS)—genotypes at thousands or millions of sites across the human genome are each tested in a sample of people for statistical association with the phenotype. Each variant identified by such a study as statistically associated with the phenotype can be assigned an effect size, representing the estimated magnitude of the increase in the trait (for quantitative phenotypes) or risk or liability for the trait (for binary phenotypes) that is associated with possession of a copy of the variant.

The effect size is measured by the correlation coefficient. However, it is called an "effect" only because it is assumed that the causal influence goes from the genotype to the phenotype. The statistical association alone does not establish that claim.

Next, these coefficients are aggregated as we saw in the case of GPS calculation:

> For many complex phenotypes, identification and analysis of contributing genomic variants—most having small phenotypic effects—has led to the

> formulation of polygenic scores, quantities that seek to predict a trait value associated with a specific genomewide set of genotypes … For a quantitative phenotype, a polygenic score for an individual genome represents an aggregation, usually in the form of a sum, of the estimated effect sizes of the genetic variants in the genome.

This means that polygenic scores inherit all the interpretive difficulties that GWAS-generated coefficients have.

These difficulties are identical to those we encountered earlier in estimating classical heritability scores. GWAS scores depend on the population from which the individuals sampled are drawn. If the environments to which this population has been exposed changes, for any complex trait, the phenotypic trait distribution would change and, therefore, so would the statistical association between trait values and individual alleles at each locus (that is, GWAS scores). If the genotypic composition of the population changes, we can expect the same type of change in GWAS scores. The emphasis here on *genotypes*, rather than allele, is intentional: for the development of a phenotype, the role of an individual allele in a genome depends on what other alleles are also part of that genome, that is, the genotype. (The other alleles are part of the *genetic environment* of any given single locus.) The parallelism between these interpretive subtleties and those encountered for classical heritability scores is exceptionless.

When we turn to the aggregation of GWAS-generated association scores to generate genomic scores, the additivity problem returns, as in the case of estimating variances for heritability scores, and even worse. Plomin and those of his ilk claim that genomic scores can be used to predict individual phenotypic outcomes. They blandly assert the association scores (correlation coefficients) represent the quantitative values of causal contributions of loci without giving any reason for that assumption. Then they weight these using the number of alleles. In each case, population-wide parameters are used to predict what is supposed to be occurring during the development of an individual. Then they aggregate by adding up. There is no biological justification for this procedure and *no evidence that it correctly predicts phenotypic values for individuals*. What we have is of as much intellectual credibility as predictive astrology, though now cast in the language of genomics.

It would take a very gullible eugenicist to believe all of Plomin's claims about the new genetics of intelligence. But even such a eugenicist would be forced to admit that the prospects of genetic enhancement of intelligence remain woefully poor at the present time even with the advances of CRISPR technology. We would have to edit scores of genes simultaneously, with all the uncertainties associate with each of them, to hope for an increase of IQ by a few points. Because each of these genes would at best have a tiny effect on the desired trait while affecting many others, the process would not come close to achieving the requirement of specificity (as developed in the last chapter in the context of editing disease-implicated genes). At the very least, designing more intelligent babies through genetic intervention is beyond our capacities right now.

WHAT ABOUT PHYSICAL TRAITS?

Intelligence, even if it is a single trait, is a complex behavioral trait. But, perhaps, the prospects for genetic enhancement become better if we stick to presumably simpler physical traits. Let us turn to a set of traits that are often viewed as embodying societal values: pigmentation of eyes, hair, and skin. All three traits depend on the presence of melanin which comes in many forms but can be classified into two types: a red-yellow form that is known as pheomelanin and a black-brown form known as eumelanin. How hard will genetic enhancement be for melanin-dependent traits?

Eye color, more specifically, the color of the iris is genetically the simplest of the three cases. Back in 1907 and 1908, Davenport and his wife, Gertrude, at the Cold Spring Harbor Laboratory, and Charles Chamberlain Hurst, a leading British geneticist of that era, claimed to have shown that eye color depends on one locus with two alleles. Though this claim finds its way into textbooks even today, we now know that it is an over-simplification. (Indeed, exceptions began to be reported as early as 1909.) Hurst recognized only two eye color phenotypes, blue and brown; in sharp contrast, a recent review points to "eye color ranges include varying shades of brown, hazel, green, blue, gray, and in rare cases, violet and red." However, Hurst was correct in one observation: brown is indeed dominant over blue or green—but even in this case there are complications.

The physical basis for eye color is the distribution and content of melanocytes producing melanin in the front layer of the iris. (Melanocytes are melanin-producing cells in the eye and skin.) A lot of melanin would absorb a lot of light resulting in a brown appearance. If there is less melanin, the color of the iris can be blue or gray or green in most cases depending on what other molecular structures are present. (The other eye colors are more complicated.) The genetics of eye color is simple in the sense that, although sixteen genes have some influence in regulating eye color, only two of them, *HERC2* and *OCA2*, both located on chromosome 15, play major roles. These genes interact with each other in such a way that, contrary to traditional accounts, two blue-eyed parents can still give birth to a brown-eyed child. In other words, the traditional story of brown eye color being dominant over blue is not strictly correct. This means, at the very least, that both would have to be edited simultaneously if we want to select a particular eye color. Another gene, *MC1R*, has been associated with green eyes and red hair though only in some populations—there will be more on this case below.

The trouble is that these two genes are not the only ones involved in eye color. Traditional genetic techniques had already identified at least nine other genes though, now, we acknowledge a role for sixteen genes. In fact, genes for any of the several proteins that play a role in melanin formation and maturation in the melanocytes also have a role in determining eye color as well as pigmentation of hair and skin. As expected, GWAS implicate scores of yet other genes and, more importantly, different ones for different regions of the world. While what that means is far from clear, the totality of the evidence indicates that no small tractable set of genes have sufficient specificity for eye color to suggest successful genetic enhancement at the present state of our knowledge; however, targeted change *may* be possible sometime in the future for embryos with very well-understood genomic backgrounds that have been carefully analyzed for the role played by *HERC2* and *OCA2*. But we do not know exactly how to analyze the genomic background with sufficient precision to guarantee the required specificity.

Human hair color is a quantitative trait that depends on the quantity, distribution, size, shape, and melanin content of organelles within cells called melanosomes. The most important factor is the ratio of eumelanin to pheomelanin. One review notes the following rules. Red and blonde hair generally contain less melanin than brown and black hair. However, blonde hair contains the same number of melanosomes as brown or black hair but has melanosomes that are smaller and rounder. Brown hair has large ellipsoid melanosomes containing mainly eumelanin. Black hair has the biggest melanosomes and most densely packed eumelanin molecules. At the same time, the review notes that visual appearance is not a very good guide to the pigment composition of hair and that hair color changes in life not only in old age but also as a child grows into an adult.

Given these complexities, it should come as no surprise that a large number of genes are involved. Of the genes we have already seen, *MC1R* is obviously relevant, but its variation seems to be important only in populations with a high frequency of red hair and fair skin. In these populations, it is associated with green eyes. Some variants are associated with freckling. These traits seem to be associated with a high pheomelanin to eumelanin ratio. As expected, GWAS data have only complicated the issue. One study of a large sample of individuals of (self-identified) European origin identified one hundred and twenty loci associated with hair color; another found more than two hundred. The first study also found a sex bias in hair color with women having lighter hair than men. Given these genetic data, genetic intervention to change hair color also remains implausible at the present state of knowledge.

We finally turn to human skin color which can vary from the darkest brown or black to the lightest white tones. Two types of skin color are usefully distinguished: constitutive and facultative. Constitutive skin color is the natural ("genetic") color of the epidermis; facultative skin color is what it becomes after exposure to ultraviolet light (for instance, by tanning) or to certain hormones (for instance, in skin-lightening creams). We will limit ourselves to constitutive skin color even though the prospect of genetically designing an individual to tan more easily raises interesting possibilities. Light skin has a high proportion of light brown eumelanin and yellow/red pheomelanins in smaller less-pigmented melanosomes that occur in bunches; dark skin has more dark brown eumelanin and larger densely pigmented melanosomes distributed singly.

Is genome editing a viable, more permanent, and even perhaps safer, alternative to skin-lightening creams? Folk biology would suggest so. Not only is skin color a quantitative trait, children typically have skin color intermediate to those of their parents: it is an almost perfect exemplar of *blending* inheritance. But folk biology is in for a surprise. The genetics of human skin color is so complicated that a 2003 review observed:

> One of the most obvious phenotypes that distinguish members of our species, differences in skin pigmentation, is also one of the most enigmatic. There is a tremendous range of human skin color in which variation can be correlated with climates, continents, and/or cultures, yet we know very little about the underlying genetic architecture. Is the number of common skin color genes closer to five, 50, or 500? Do gain- and loss-of-function alleles for a small set of genes give rise to phenotypes at opposite ends of the pigmentary spectrum? Has the effect of natural selection on similar pigmentation phenotypes proceeded independently via similar pathways? And, finally, should we care about the genetics of human pigmentation if it is only skin-deep?

Fifteen years later, we are not doing much better.

From a medical perspective, we should care about human skin color insofar as different amounts on melanin can, on the one hand, influence susceptibility to cancer induced by ultraviolet radiation and, on the other, affect the ability to synthesize vitamin D and prevent rickets. The genetics of coloration is complicated. Though smaller numbers of genes have been implicated than, say, for hair color, different genes have been implicated for different populations. A 2007 study of South Asians found three genes—unhelpfully named *SLC24A5*, *SLC45A2*, and *TYR*—to be most important in explaining variation in skin tone. For a European population, the six genes implicated were *HERC2*, *OCA2*, *IRF4*, *TYR*, *ASIP*, and *MC1R*; only *TYR* was shared with the

South Asian population but recall that *HERC2* and *OCA2* both play major roles in regulating eye color.

A study of African populations from Ethiopia, Tanzania, and Botswana implicated *SLC24A5*, *HERC2*, and *OCA2*, as well as three other genes (*MFSD12*, *DDB1*, and *TMEM138*). A study of KhoeSan populations in southern Africa found none of these genes important for skin color variation. A study of Latin populations implicated *TYR*, *OCA2*, *HERC2*, *SLC24A5*, *SLC45A2*, and *IRF4*, but also *TYRP1* and a new gene not previously known to be associated with skin color. At the present time, germline editing would be no viable competition for skin-lightening creams.

GENETIC REDUCTIONISM

What we have just seen is that genetic enhancement has fallen afoul of the failure of a doctrine that philosophers have dubbed *genetic reductionism* and we have encountered this doctrine several times earlier in this book during discussions of what drove the HGP. Since this failure may well doom the project of liberal and moderate eugenics, at least for the foreseeable future, the point will be worth elaborating in some detail. For philosophers, *reductionism* in general is the doctrine that laws and facts at one level can be explained by (and, in that sense, *reduced to*) laws and facts at another more fundamental level. Suppose that all biological phenomena can be explained by ordinary physical and chemical interactions. This is the thesis of *physical reductionism* that was so enthusiastically promoted by Pauling in the 1950s as we saw in the second chapter.

We also saw there that molecular biology since the mid-twentieth century provides many interesting examples of successes of physical reductionism including the operon model for enzymatic adaptation and the allostery model for the Bohr effect. In both cases, puzzling (and important) phenomena from cellular biology were reduced to macromolecular physics. Indeed, many philosophers have viewed the emergence of molecular biology as a triumph of such physical reductionism though this interpretation remains controversial.

Physical reductionism must be contrasted with a related but fundamentally different research program in twentieth-century biology, that of *genetic* reductionism. The motivation for this program was the belief that genes were the most important determinants of all aspects of an organism's structure and functioning, that is, all its phenotypic traits. In the 1920s, this view seemed plausible enough when geneticists were systematically showing that large numbers of traits were inherited as predicted by Mendel's laws of genetics. In particular, as we saw in the first chapter, in the famous fly room at Columbia University, Morgan and his students showed

Mendelian patterns in the inheritance of hundreds of traits in the fruit fly *Drosophila melanogaster*.

On the basis of these observations, it was believed that genes were causally responsible for the traits even though, unlike the situation of physical reductionism in molecular biology, the causal pathways remained unknown. At the molecular level, even today, these pathways have only been fully worked out for a handful of traits. We have every reason to believe that the physics and chemistry of molecules, starting with DNA, proteins, and other molecular types in cells, ultimately lead to the most complex phenotypes of organisms. But we remain far from being able to compute the process, step by step. Even though we know that the production of phenotypes takes a wide class of molecules, and not only DNA, genes still remain the focus of much of the research because DNA is the factor we know best how to manipulate in laboratories. Thus, genetic reductionism has become somewhat like a null model for much of biological research even though its limitations are widely recognized by biologists, if not psychologists.

Genetic reductionism is the program that lies behind the claims of Plomin and others who are willing to claim genetic causality without any knowledge of the causal pathways from the DNA molecules that specify the genes to the trait itself. Genetic reductionism does not concern itself with the molecular mechanisms by which a trait emerges in an embryo. In the heyday of genetic reductionism in the 1920s, genes were supposed to be proteins rather than DNA. Yet, this did not prevent confident assertions of genetic causality: that is how distant genetic reductionism has always been from the exploration of causes. We are not suggesting that genes do not play an important causal role in the emergence of traits. What we are pointing out the fragility of any assumption that genes are the sole (or even the most) important causal determinants of all traits.

When the HGP was envisioned and promoted in the biological community in the 1980s, critics pointed out the extent to which it was based on the assumption of genetic reductionism. The HGP was supposed to deliver a complete human DNA sequence, eventually a unique one for each person. This sequence would provide a person's genetic profile (assuming that we can fully work out which parts of the sequence are genes rather than junk DNA). Now, the expectation was that a knowledge of these profiles would revolutionize biology and medicine. We have dealt with this story in the third chapter. What genomics has shown is that the promised revolution of the HGP must be postponed, perhaps indefinitely so.

To put it bluntly: genetic reductionism is a failed doctrine that should be relegated to the dustbin of history. There are a few hundred (organism-level) human traits that are largely specified by one or two genes. These are well known because they have been easy to study using genetical experiments and because geneticists in the twentieth century were always looking for them because they were easy to study. (The same situation holds for other species.) But, once we turn to other traits, typically ones that are interesting for ordinary reasons, including eye, hair, and skin pigmentation, we find that a large number of genes are involved and that they interact with an array of nongenetic factors. The history of these interactions matters.

Contrary to the expectations of those who pushed the HGP in the 1980a, the genomics research spawned by the HGP has only demonstrated the limitations of genetic reductionism. Knowing genetic profiles provides very little insight into the functioning of an organism and is only very occasionally relevant to medicine. (Moreover, where they are relevant, as in the cases of disease genes discussed in the last chapter, the role of these genes was known long before the HGP.) The rise of epigenetics is one of the understandable results of recognizing the limitations of genetics: if genes alone explain very little of the biology of the organism it is time to move to other molecules present on chromosomes or elsewhere in the cell. Even GWAS results provide reason for increasing skepticism about how important genes are. When hundreds of genes are associated with a trait (or, typically, with variation in a trait), but each has a tiny nonspecific effect, knowing only these genes tells us very little about causal pathways, how an organism ends up with a trait during the course of its life. We must know how these genes are used, in what sequence, and what environmental factors must be using these genes.

CONTEXTUAL DEVELOPMENTAL CONSTRUCTION

Thus, the limitations of what germline editing can achieve in terms of genetic enhancement are biological rather than technological. These limitations are not confined to CRISPR technology as it exists today; it applies equally to technologies of editing that will likely emerge through improvements to current CRISPR methods and even to better gene-editing techniques that may supersede all the ones that exist or are envisioned today. Genes alone do not make an organism.

Rather, a mature organism is the result of a sequence of developmental interactions between an embryo, starting as a single cell, and its environment. This view of life is called *contextual developmental construction*. In sexually reproducing organisms, within the fertilized cell, expression of genes depends on a variety of epigenetic factors. This is true of all organisms, and not just humans. (In asexual organisms we start with a single cell starting the reproductive cycle by initiating development.) The physical environment matters tremendously, and not just because proper development requires the presence of water and a suite of chemical nutrients. Sometimes, even gravity matters.

In experiments first carried out in the 1970s and 1980s, developmental biologists showed that, if chick or frog embryos are tilted to an incorrect angle in early development, the body does not have the normal shape and symmetry. Recent work in animal development has also emphasized the importance symbionts living on and in the body—our gut bacteria make differences to our developmental pathways.

For some traits, notably sex determination, temperature is the most important variable in several species. For instance, the sex of most turtles and all crocodilians (alligators, crocodiles, caimans) is determined by the temperature at which eggs are incubated. Typically, only a very small range of temperatures allows the formation of both males and females. Above (or below) that range only one of the sexes is formed. In most cases, lower temperatures lead to males and higher ones to females. But temperature is only one of the environmental determinants seen for sex in animals. In *Gannarus duebeni*, a tiny amphipod (a type of crustacean), sex is determined by the time of exposure to sunlight: young which are born earlier in the season turn mostly male, the later-born mostly turn female. Perhaps even stranger is the case of the worm, *Bonellia viridis*: if the larvae settle on the seafloor, they become female; however, if they are ingested by a female they migrate to the uterus and becomes male.

Temperature-dependent sex determination is an example of phenotypic plasticity: when organisms with the same genotype develop different phenotypes in different environments. This aspect of development was recognized shortly after the beginnings of genetics in the early twentieth century. Phenotypic plasticity is ubiquitous. The European map butterfly, *Araschnia levana*, has a spring form that is bright orange with black spots and a summer form that is almost black with a white band. The forms are so different that Linnaeus classified them as different species back in the eighteenth century. The water flea, *Daphnia cucullata*, grow their "helmets" (parts of their head) to twice the normal size in the presence of predators. The induced change is inherited by the next generation. In humans, phenotypic plasticity is shown (in a perhaps trivial sense) by every cultural trait when there can be variation independent of genetic changes. It has also been interpreted as the capacity of cells to change their behavior in response to internal or external environmental cues.

Phenotypic plasticity shows the extent to which development depends on environmental context. It shows how organismic forms—structure and behavior—are differently constructed in different environmental contexts using genetic capacities in different ways. If a single gene is being analyzed, its environmental context consists not only of extra-genetic factors but also of the presence of other genes in that organism's genome. The existence of potential phenotypic plasticity underscores the importance of ensuring specificity before embarking on a program of gene editing to achieve desired phenotypic changes.

WHAT IS PERFECTION?

We started this chapter with a 1932 exchange between J. B. S. Haldane and Francis Albert Eley Crew, then Buchanan chair of Animal Genetics at the University of Edinburgh and one of the more prominent geneticists of that era. The occasion for the exchange was the Sixth International Congress of Genetics being held at Cornell University in Ithaca, New York, during which Haldane was being interviewed for his views on eugenics by William L. Laurence of the *New York Times*. The interview had coalesced around the concept of a genetically "perfect man" when Crew happened to walk by and was interrupted by Haldane. Crew's quip has ended up in history books as the perfect response to naive eugenic claims. But what it means is less than clear.

At the time of the interview, Haldane was no principled opponent of eugenics though his skepticism about many eugenic measures slowly strengthened through the 1930s primarily due to Nazi abuses of genetics. Moreover, Haldane defended one eugenic proposal throughout his life: the use of non-coercive eugenic measures to combat and eliminate genetic diseases. Even here, Haldane was much more conscious than his contemporaries about how difficult elimination would be in the case of recessive diseases: symptomless carriers of a disease gene would often not be recognized and would pass on such genes to future generations. Nevertheless, negative eugenics was fine, and socially desirable. Haldane's perspective was not much different from the one we suggested in the last chapter. On this front, not much has changed after eighty years.

But when it came to positive eugenics, what is now called genetic enhancement, Haldane and Crew were vocal skeptics but for reasons rather different from the ones we have just discussed. Both of them had much more confidence of the role of genes in traits connected with intellect or temperate. In the 1930s, this confidence was more reasonable than it would be now. After all, during the 1920s, as we have seen, work by Morgan's laboratory had established the role of genes in the development of hundreds of traits in fruit flies. Work on other organisms, though not as detailed as the focus on fruit flies also supported a central role for genes. Moreover, Haldane and Crew were obviously not privy to all the results that have emerged in the last few decades to underscore the complexity of the relationship between genes and traits.

Rather, Haldane argued—and Crew concurred—that there is objectively no such thing as the "perfect man." It depends on the perspective, as we noted earlier in this chapter. Haldane thought that eugenicists demanding human improvement were merely promoting what society valued as desirable at the expense of individual liberty and individual differences. They were pandering to societal prejudices (which was one of our worries about genetic enhancement). He was appalled by involuntary sterilization in the United States and, presciently, by the rise of the Nazis in Germany. Crew underscored this role of society in his quip: angels are angels only in the context of what the values of a society happen to be at the time: they are the ones that exhibit these values most prominently (presumably with the sole exception of God but, then, God wouldn't be part of society in any case).

Haldane and Crew saw little of value in positive eugenics. Instead of individual enhancement to fit better in society, Haldane suggested that societies set their goals to accommodate the differences and desires of the individuals who make them up. Much of the interview consisted of Haldane emphasizing the value of diversity with society: without diversity, society would not be functional let alone interesting. Writing in the 2020s, the perspective seems eerily contemporary. Haldane was ahead of his time, as he often was on both biological and social issues. But, if a good society consisted of a diverse array of individuals, there is no "perfect man," no defensible eugenic idea of betterment to perfection, no value in attempted genetic enhancement. Even leaving the scientific problems aside, there is no plausible ethical rationale for genetic enhancement and abundant ground for skepticism. Eugenics should stop with the elimination of the disease. Some eighty years later, we would do well to listen to Haldane.

Chapter 8

A CRISPR Future

> "The Model T was cheap and reliable, and before long everybody had a car and the world changed . . . CRISPR has made gene editing cheap, easy and accessible, and therefore more common. I think it's going to change the world. *Exactly how beats me*."
>
> —Hank Greely, 2018, in Schwartz, M., "Target, delete, repair."

EDITING THE HUMAN GERMLINE

If the biological discussion at the end of the last chapter is correct, as far as the prospects of human genetic enhancement of complex traits are concerned, the CRISPR bubble will eventually burst. There will be no designer babies. Assuming that insurmountable future problems do not arise and future development of CRISPR techniques resolves issues such as specificity, CRISPR's main eugenic contribution will be to remove a number of disease-causing genes from the human population. We are in no position to estimate how many such genes there might be. In a few cases, such as myotonic dystrophy, the argument for elimination remains simple and convincing. In general, whether a gene should be eliminated must be a societal decision at the global level, and must not be treated as a purely technological issue. We should also not forget that CRISPR-based gene editing allows for the easy elimination of a gene without aborting a fetus. Thus, it avoids the critical ethical problem that opponents of abortion have had with traditional eugenic measures based on prenatal genetic testing and removal that envision abortion of a fetus or, at least, destruction of an embryo.

For myotonic dystrophy, there is no good reason to allow its gene (that is, its allele) to persist in the human population. This allele presents a straightforward choice because it is dominant (a single copy suffices for the disease to manifest itself) and has high specificity. Moreover, the disease is severe and it is implausible that there will be advocacy groups that would include people carrying the gene who support its continued persistence. This disease is also relatively common and, therefore, among those that are most important in a medical context. Elimination will likely also be a simple choice for rarer disease-implicated genes that are dominant and have high specificity. For instance, most cases of polycystic kidney disease are the result of a dominant mutation in any one of three genes. Each of these mutations has high specificity and probably can—and should—be targeted for elimination without undue controversy.

Once we turn to more complex genetic diseases, the situation gets murkier but, in some of these cases, elimination may well come to be viewed as the wisest social policy. For instance, the heterogeneity of Huntington's disease mutations raises questions about *when* to prescribe elimination though not about whether the disease gene should be eliminated. Perhaps, a wise policy would be to slate the gene for elimination once the number of CAG repeats crosses the higher threshold of normalcy at twenty-six repeats. But this is a matter that should be discussed as a question of social policy.

What remains troubling right now is that though there have been multiple committees and working groups set up by various national academies and other august bodies in multiple countries to study the ethics of gene editing, these groups have continued to discuss and debate abstract questions about the ethics and desirability of germline and other gene editing, but the discussions have not focused on the practical questions that we are already beginning to face: the which, when, and where of gene editing and elimination. For instance, there is no compelling reason to delay the elimination of the mutation for myotonic dystrophy from the human germline provided that the safety of the CRISPR-based (or any other) procedure is adequately established. But none of these committees or working groups have had the courage to make such obvious recommendations. Nor has any of them provided reasons for not acting. The only partial exception is the recent report from the US National Academy of Sciences. It breaks new ground by accepting that single-gene diseases are the most pertinent targets, as we have also suggested in this book. But, then it claims that the safety of CRISPR technology has not been sufficiently established. In this, the report may well be correct (as we have also seen in this book) but it did not lay out specific criteria for safety that must be satisfied before recommending germline editing.

As this report implicitly admits, elimination is an attractive option for genes for several single-gene recessive diseases including sickle cell disease and hemophilia. For both, a carrier who is heterozygous for the trait will not suffer from the disease. In the case of hemophilia, the gene responsible for the disease is located on the *X* chromosome. So, whereas men with such an *X* chromosome will suffer from the disease, it is extremely unlikely that a woman would suffer from the disease because the hemophilia gene, which is rare, would have to be present on both her chromosomes. Yet, a heterozygous woman would pass that gene to half her children on the average. Should such genes be eliminated?

This discussion has been framed as if the decisions to be made will be a matter of social policy. However, in the spirit of liberal eugenics, it accepts that social policies will consist of offering strong recommendations, leaving the final decisions to parents. The assumption that social policy will play the determinative role is likely to be correct at least for the foreseeable future worldwide because many reproductive decisions such as abortion continue to be regulated through social policy. In some countries, such as India, there are even legal restrictions on testing fetuses for sex. Yet, we should also keep in mind that some other similar reproductive decisions are left to parents (along with their medical advisers), for instance, during the screening of embryos for implantation in the uterus as part of *in vitro* fertilization procedures.

Parents are allowed—indeed, often even encouraged—to reject genes associated with serious diseases. Genetic conditions that can be screened in this way include all the diseases that have been discussed in this book. So, the question becomes: if parents are permitted to use genetic screening to reject embryos with certain genes after *in vitro* fertilization and before implantation, why should they not also be permitted to edit those very same genes in embryos before implantation? If we should decide to prevent such an outcome, the policy decisions we make better be backed by arguments that are convincing enough to generate wide societal consensus.

GENE DRIVES

Germline editing of humans against genetic diseases (or, potentially though much less plausibly, to enhance traits) is not the only use of CRISPR technology with vast social consequences. Gene drives against species perceived to be undesirable are an equally potent application. A gene drive is a mechanism by which a gene is transmitted across generations in such a way that it spreads faster through a population than allowed by Mendel's laws. A gene drive begins by creating a molecular assemblage called a gene drive *construct* that consists of a guide RNA designed to target an intended DNA sequence, a gene for the enzyme Cas9 that will cut the DNA at this target, along with a "cargo" of genes to be inserted at the target. (Cas9 can be replaced by other Cas enzymes that also cut DNA.)

There is no restriction on what genes can be included in the cargo and, in a standard gene drive, the cargo includes so-called selfish DNA that copies itself wherever it can in the genome. The designed construct and Cas9 enzyme are introduced in the cell. This Cas9 cleaves the targeted DNA in one of the two homologous chromosomes and the construct gets inserted into it. The inserted genes are designed to now target the homologous chromosome which is cut by Cas9 at the corresponding site. The construct then gets copied into the second chromosome through DNA repair processes. Cas9 continues to be produced as necessary because a gene for it is part of the construct.

Thus, after a gene drive construct is inserted into one chromosome in a cell, both homologous chromosomes in the cell will end up the cargo of inserted genes. When this cell is a precursor of a germinal cell, it contributes two copies of the construct to the next generation. In those descendant organisms, the inserted construct then targets the homologous chromosome and inserts itself into it. As this process iterates generation by generation, the construct including its cargo of genes rapidly spreads through a population instead of reaching a stable frequency as would happen with ordinary Mendelian inheritance.

Gene drives can potentially be used to control and eliminate vectors of disease such as insects or ticks, agricultural pests, as well as invasive species. In gene drives designed for the suppression of a species, the inserted DNA would be designed to reduce the mean fitness of populations; for instance, by disrupting a gene that affects essential functions. The size of such a population would decline over generations. In the case of insect vectors of disease, gene drives aimed only at disease control can also be designed to prevent pathogen transmission by the vector. We will discuss one such example below, the mosquito, *Anopheles stephensi*, which transmits malaria.

The possibility of extirpating populations of undesirable species using CRISPR-based gene drives has been receiving increasing attention since 2015 though the idea of a drive using selfish DNA for this purpose goes back to Austin Burt of Imperial College London in 2003. Four developments in 2015 raised the prospect of eliminating harmful insects through gene drives. Ethan Bier and Valentino Gantz of the University of California at San Diego published a pioneering study of a CRISPR-based system for a gene drive. They targeted the recessive yellow (*y*) gene on the X chromosome of the fruit fly *Drosophila melanogaster* that makes the flies turn completely yellow. Their gene drive construct used a Cas9 gene and a guide RNA flanked on each side by regions that recognized the targeted gene. When they crossed females with this construct to normal males, 95–100 percent of the next generation seemed to contain the construct rather than just 50 percent as expected from Mendelian inheritance. Though this percentage later turned out to be an over-estimate it was clear that the frequency of the inserted construct was much higher than 50 percent. These numbers provided a credible demonstration of how CRISPR-based gene drives can be used to spread genes rapidly through a population.

Bier began a collaboration with Anthony James of the University of California at Irvine who had previously isolated genes that make mosquitoes resistant to the major malarial parasite, *Plasmodium falciparum*. Their collaboration resulted in a successful deployment of a gene drive in a laboratory population of the mosquito *Anopheles stephensi*, which is an important vector for spreading malaria in South Asia. They reported a 99.5 percent transmission of the resistant genes to the next generation. The experiments also showed transcription of the resistant genes in the offspring. Though they emphasized that gene drives alone would not eradicate malaria and underscored the value of "therapeutic drugs, vaccines, and alternate vector-control measures," the successful deployment of these gene drives was viewed as a major step toward the eventual control of malaria.

Even more hope—and hype—was generated from Imperial College in London. A team led by Tony Nolan and Andrea Crisanti engineered a gene drive in laboratory populations of *Anopheles gambiae*, the most important vector of malaria in sub-Saharan Africa. They identified three different genes that halted egg production in females when disrupted. They then inserted a gene drive construct into each of these genes, and confirmed the transmission of the construct to the next generation. For one of these genes, they carried out experiments on mosquitoes confined to laboratory cages in which reduced fertility spread to 70 percent of the population over four generations.

Finally, Zach Adelman and colleagues at Virginia Tech identified a dominant gene that causes maleness in *Aedes aegypti*, the most important mosquito vector for dengue, chikungunya, yellow fever, and Zika. They showed how a CRISPR-based system could be used to insert this gene into the genomes of female mosquitoes to convert them to males. Male mosquitoes do not need blood meals; because of that they do not bite humans or spread disease. Moreover, the spread of maleness in a population at the expense of females is a precursor to its eventual extirpation. These results raised the prospect of designing and using gene drives against *Aedes aegypti*. Given the extent to which dengue had spread during the preceding three decades (more so than any other insect-borne disease, according to the World Health Organization) and the fear of Zika in the ongoing epidemic at the time, using this strategy of disease control had many proponents in 2015.

Further experiments after these initial results have somewhat dented the early hopes of gene drives. The first challenges emerged in 2017. An array of theoretical analyses and experimental results highlighted the technical limitations of CRISPR-based gene drives in insects. Evolution limits what gene drives can easily do. The first limitation comes from the basic fact that all natural populations have genetic variation that can both prevent recognition of the intended target by the RNA guide of the gene drive construct and also foil cleavage by Cas9 (or any other substitute enzyme introduced to cut the DNA). Moreover, evolved resistance can arise through various mutational mechanisms that would be selected for if the drive reduces fitness as, for instance, was the case in the Imperial College experiments that disrupted fertility in female mosquitoes as we will see below. Breeding patterns other than random mating can also slow the spread of a gene drive in a population. If, for instance, wild organisms tended to mate much more with other wild ones, rather than with laboratory-released ones carrying the gene drive construct, the inserted DNA would not spread very rapidly. There is some evidence of this type of mating preference in mosquitoes.

A group at Cornell University began experimentally studying the emergence and evolution of resistance to gene drives in the fruit fly *Drosophila melanogaster*. In a 2017 paper, they showed that resistance to gene drives arose easily even in the absence of selection and within one generation. Their results suggested that resistance arose as a result of how Cas9 acts. After cutting by Cas9, instead of being converted to genes to propagate the drive, the DNA was being "misrepaired" into what came to be called "resistance alleles." Later in 2017, the Imperial College team also reported the emergence of resistance alleles. They had been optimistic in 2015 when their first experiments with *Anopheles gambiae* had shown the rapid spread of the gene drive over four generations. However, when the team allowed the experiment to run for twenty-five generations, the spread halted after six generations and the frequency of the drive allele in the population rapidly decreased afterward. Their results pointed to natural selection for resistance to the gene drive. The implication of this result was clear: it will be much more difficult to use gene drives to control malaria in sub-Saharan Africa than what was believed in 2015.

Meanwhile, a team at Indiana University analyzed the potential effect of natural genetic variation on CRISPR-based gene drives in the flour beetle *Tribolium castaneum*, an insect pest that is estimated to consume a fifth of the grain produced in the world each year. Their results showed how small levels of genetic variation, especially when accompanied by some inbreeding, could prevent gene drives from decreasing population sizes. They analyzed variability at three drive-relevant genes in four populations of flour beetles (from India, Peru, Spain, and Indiana). Using available sequence data, they looked at PAM sequences of DNA in the targeted gene that are essential for cleavage by Cas9. They found that most populations harbor natural variants at such Cas9-relevant sites in sufficient frequencies to prevent gene drives from propagating successfully.

These difficulties are not insurmountable and gene drive researchers have proposed many ways out. The simplest is to target multiple functional genes for disruptive editing simultaneously. Even if existing variation or evolved resistance rescues one gene from being successfully targeted, it is highly improbable that all targeted genes would be simultaneously rescued in this way. In 2018, the Cornell University team showed that the use of two different genes as targets reduced the evolution of resistance alleles enough to allow a gene drive to spread through a population of fruit flies.

THE ETHICS OF PROMOTING EXTINCTION

Gene drives come with their own ethical dilemmas. Start with a drive that spreads a fatal gene across an unwanted species. The basic gene drive mechanism described earlier would facilitate the spread of that gene across the entire species unless the targeted populations into which the genes are intentionally introduced are kept completely isolated from all other populations. This level of isolation is more or less impossible to ensure in practice which means that the species as a whole could be driven to extinction. Most environmental advocates would find such an outcome unpalatable: we would have done more than just allow a species to become extinct. We would have *knowingly* taken an action that was likely to drive it to extinction, that is, we would have facilitated extinction.

Such a risk of extinction is probably highest when gene drives target populations of invasive species for extirpation because these species are typically very mobile. Very often it is this ability to disperse easily that is responsible for them to have expanded their range to such an extent that they come to be regarded as invasive species in areas outside their original range. The trouble is that a gene drive introduced at a place where the species is invasive could easily spread back to populations of that species in their native range and also drive them extinct.

There are additional potential unintended consequences. Not all species are fully reproductively isolated from each other, that is, sometimes individuals from different species mate to produce fertile hybrids. This provides a simple way for a gene drive to jump from its intended species to another. Gene drives could also spread from one species to another through a variety of mechanisms of what is called horizontal gene transfer, that is genes that move from one individual to another by some form other than through reproduction. Horizontal gene transfer between species can occur through viruses, parasites, symbionts, and other entities that move from one species to another. Some cases are eerily similar to gene drives because they rely on selfish DNA elements. For instance, a string of DNA called the *P*-element has invaded populations of the fruit fly, *Drosophila melanogaster*, worldwide after a single horizontal transfer from another fruit fly species, *Drosophila willistoni*. It is now also spreading through *Drosophila simulans* probably by hybridization with *Drosophila melanogaster*.

Extinctions are bad enough but even the extirpation of a local isolated population may have unwanted (and unintended) ecological consequences. For instance, gene drives against rodents have been proposed quite seriously, at least for islands. It is a fact that rodents threaten the endemic biodiversity of most islands in the world today and are considered invasive species that have high priority for eradication with or without gene drives. The rodents that do most harm are three rat species: the house or black rat (*Rattus rattus*), the brown rat (*Rattus norvegicus*), and the Polynesian rat (*Rattus exulans*); as well as one mouse species, the house mouse (*Mus musculus*). In the case of mice, one plausible idea is to introduce a gene that severely distorts the sex ratio in favor of males thus decreasing population size each generation.

However, in many islands, these rodent species have been present for so long that they have become functional components of the ecosystem. Simulation studies have shown that their rapid eradication would require repeated introductions of gene drives and the speed with which this is accomplished increases the severity of ecological disruptions. Most of these disruptions are likely to harm other species on the islands. Thus, even when extinction of a species is not at stake there may be good reason to be cautious about the introduction of gene drives in natural populations. But it is also the case that conventional methods of rodent eradication have harmful ecological consequences especially when poisons are used and can potentially be ingested by other species. There is no perfect course of action.

In some cases, we may even be open to the possibility of extinction of a species. Consider, for example, the yellow fever mosquito, *Aedes aegypti*, or the Asian tiger mosquito, *Aedes albopictus*. Between them, they spread chikungunya, dengue, yellow fever, and Zika, along with a variety of rarer diseases. In spite of many attempts at elimination, yellow fever continues to exist in parts of Africa and South America. For the past fifty years, dengue has been the fastest spreading insect-borne disease and its reach is increasingly becoming global. Chikungunya has already become a major problem in parts of Africa and Asia and is also expanding its range. We went through a global Zika crisis in 2015. Without these Aedes species (and a few other far less important ones from the same genus), these diseases cannot spread.

So, why not drive these species extinct? There are about three hundred mosquito species in the world and not one of them individually or all of them collectively are known to be a major part of any food chain in any ecosystem anywhere. We also have plenty of experience with regions in which mosquito species have been eradicated to control diseases such as dengue and malaria.

For instance, the insect vector of malaria, *Anopheles gambiae*, was eliminated from northern Brazil in the late 1930s and the yellow fever mosquito was eliminated from parts of Florida in the 1950s. The yellow fever mosquito was eliminated and disappeared from almost all of the American continent for most of the 1960s though it has reestablished itself in tropical South America since the early 1970s after the use of DDT as a pesticide was banned. In no case has there been any semblance of ecological collapse. There is no good ecological reason to fear mosquito extinction.

Yet, there will be many who will find any attempt to drive a species to extinction unpalatable. What we are encountering here is a manifestation of the Noah Principle, a deeply ingrained societal norm in the United States and much of the North. This is the biblical admonition that God is supposed to have conveyed to Noah when he was asked to preserve a mating pair of every animal before the Flood. In the United States, it is reflected by the most important piece of legislation for biodiversity conservation ever enacted, the Endangered Species Act of 1973 (even though that law specifically excludes pest species). So, we are faced with a social choice: is the preservation of every species so important that it trumps the eradication of deadly disease like malaria and dengue?

At the end of the day, though, worries about species extinction may be fanciful. Biologists have proposed a wide array of technical solutions to the problem of keeping gene drives under control or even reversing them. Gene drives can be reversed by introducing a new drive that disrupts the functioning of the first construct. All that would be needed would be a CRISPR-based construct that targets the first one. However, there is one caveat: the genome would still be carrying the detritus that would remain from the first insertion. Though we would have restored the original functioning of the organism, we would not have recovered the original genome.

Controlling the spread of a drive is even more interesting. One intriguing idea is to develop *daisy-chain* gene drives that are designed to dissipate over time. In these, the CRISPR gene drive construct is broken up into a number of different elements each of which inserted into a different region of the targeted genome. Suppose there are three such parts, *A*, *B*, and *C*. By design, *A* does not drive itself but drives *B*, that is, *A* does not spread by copying itself but enables B to be copied. *B*, in turn, does not drive itself but drives *C* which produces the wanted effect in the population, for instance, by turning all organisms male.

Suppose that a certain number of *A* elements are introduced in the population. As long as they are present, the frequency of *B* continues to increase and that of *C* increases even more. But, because selection acts against *A*, it eventually disappears and the drive grinds to a halt. This means that the desired trait spreads through the population for a while but the spread is self-limiting.

Such a drive could be made more and more powerful by making the daisy chain longer. But, in the end, it would not spread uncontrolled through an entire species so as to drive it to extinction.

When it comes to preventing extinction, these complicated gene drives may turn out to be entirely unnecessary. Recall all the problems biologists have had to maintain a gene drive spreading through a mosquito population in a laboratory, so much so that deploying a gene drive against malarial mosquitoes in the field still remains a pipe dream. Species targeted for being important disease vectors or invasives are likely to have large natural geographically widespread populations. (Otherwise, they would not be a perceived problem.) Such species are likely to have large amounts of natural variation much of which could confer resistance to harmful gene drives. Resistance, even to multiple harmful genes, is also likely to evolve easily and in a variety of forms in such species. Extinction is so highly improbable that there seems to be little reason to delay the deployment of gene drives against species such as *Aedes aegypti* or *Aedes albopictus* so long as all other problems such as safety are properly resolved.

Even then, as the University of California researchers pointed out, gene drives alone will not be enough to control, let alone eradicate, diseases such as malaria or dengue. The most likely scenario is that they will help drive the size of insect populations down sufficiently so that, when used with other measures such as traditional pest control, along with advances in medical strategies such as vaccination and therapy, these measures together may finally eradicate these diseases. Moreover, as it is likely that new forms of resistance to the CRISPR constructs will continue to evolve, new gene drives must be conducted in what would be a continuing struggle against vector-borne diseases. CRISPR may help tremendously, but it is not a panacea for all problems.

BIOSECURITY

CRISPR technology has security implications. The credible threat is not that state actors or rogue organizations will use germline editing to design super-soldiers though this worry has been raised time and again. In August 2018, I was questioned—quite seriously—by an FBI agent in Austin, Texas about this possibility. I believe I was able to assuage all immediate fears. The discussion of the last chapter shows any such possibility is beyond the reach of any gene-based technology that we know of today and likely impossible because of the fundamental biology of development.

But there is always the possibility that gene editing will be used to make more lethal versions of disease agents such as viruses or bacteria. This possibility has been around since the dawn of recombinant DNA technology in the 1980s. As Douglas Feith, US deputy assistant secretary of Defense, put it in the *Washington Post* in 1986: “the stunning advances over the last five to 10 years in the field of biotechnology . . . mean new and better biological weapons for any country willing to violate the international norm against the possession of such weapons.” Feith went on to worry about the development of “designer biological weapons.” In retrospect, these worries also seem unduly exaggerated. As far as is publicly known, no nation has successfully developed biological weapons through recombinant DNA.

CRISPR technology does make designing biological weapons a lot easier than before and security establishments have duly taken notice. In the United States, the Pentagon and the FBI has recently been active to identify and investigate CRISPR-based threats. If all CRISPR technology did was make editing genes easier, there would probably be little reason to suspect that it would be much more dangerous that the earlier recombinant DNA techniques. What makes CRISPR-based risks new is the possibility of using gene drives and that is what the security establishment has mainly been interested in. There is at least some reason to worry. If CRISPR-based gene editing only makes the design of biological weapons easier than in the 1980s, gene drives make their deployment much easier than anything envisioned in any previous period.

In 2016, a team of undergraduate students from the University of Minnesota attempted to design a gene drive as a project to enter into the annual iGEM (International Genetically Engineered Machine) competition that brings together student researchers from around the world. The project, “Shifting Gene Drives into Reverse: Now Mosquitoes are the Yeast of Our Worries,” was not to engineer a new gene drive to spread in yeast. Rather, the team’s concern was the safety of gene drives and how they could be reversed if they turned out to be harmful. So they attempted to design and introduce a gene drive construct that would halt an existing drive. They did not succeed. But, to the alarm of the competition’s administrators and others, they came surprisingly close.

The team correctly identified the DNA sequences they needed. iGEM allows competing teams to obtain many components of projects for free but that did not include the entire sequences that these undergraduates needed. So, they divided up the sequences into smaller parts and got each of them from a different supplier before trying to synthesize the final products. Though they managed to assemble some of the components, they never succeeded in resolving contamination issues and suffered other setbacks because the fridge they used was not cold enough to keep the reactants stable. With a little more time, a little more experience in working cleanly with DNA, and a better fridge, they may well have succeeded.

The risk of “rogue” gene drives comes from the low cost and simplicity of CRISPR-based gene editing and the emergence of a growing international ("do-it-yourself") DIY Bio movement. This movement is global in its scope and has its own organizations with many practitioners coordinated by the DIYBIo.org organization founded in 2008. Members may work individually or in community laboratories. Cities that have well-established communities include New York, Baltimore, and San Francisco in the United States as well as Manchester in Britain; there are groups in New Delhi (India), Tel Aviv (Israel), and São Paulo (Brazil). Right now, DIY CRISPR kits for bacteria can be bought over the internet for about US$ 150.

Biosecurity experts continue to fret about CRISPR experiments in garages that could go awry. While such safety issues should not be belittled, the DIY Bio community as well as iGEM have been active in monitoring the safety of CRISPR research by amateurs. Moreover, as the experience of the University of Minnesota undergraduates shows, CRISPR-based gene editing for gene drives is not trivial. It will be a little while before individual or small groups of dedicated amateurs have the knowhow and facilities to create even a partly functional gene drive. But accidents can happen and spread beyond the intended target especially when experiments are carried out beyond the containment of secure laboratories.

The real security worry is the potential for the intentional harmful use of gene drives by rogue operators, whether they be individuals, extremist groups, or state-sponsored agencies throughout the world. At present, individuals may not be in a position to design effective gene drives by themselves. While we do not know—at least on the basis of publicly available information—whether any extremist group has facilities conducive for CRISPR research, it does not appear credible. If these groups have established sufficient competence in molecular biology to pose a threat for us, it is likely that we would already have seen attempts at biological warfare against us. Instead, as Kathleen Vogel, and Sonia Ben Ouagrham-Gormley have put it: "we have little empirical data over the past 30 years that show a specific state or terrorist group using any of these new biotechnological innovations to create biological weapons."

Nevertheless, state-sponsored agencies are probably the greatest threat. For many countries, these entities would have the minimal resources needed to create a molecular construct for a gene drive involving some wild species. Since the intent is to do harm, such an actor would not have to worry about containment in case the deployment went wrong. Moreover, even if it went wrong, it is unlikely that the attempt would be detected because no wild species (except perhaps some critically endangered ones) are ever monitored that closely.

In one scenario, imagine introducing a gene drive construct in Aedes mosquitoes that make them resistant to the latest pesticides developed against them. If this construct is designed for the yellow fever mosquito, a species does not spread very easily, a particular region could be targeted for the spread of diseases such as dengue and Zika. If the intent is to target a larger geographical region, the construct could also be introduced in the Asian tiger mosquito which rapidly expands its range. If the first attempts fail, even in as economically advantaged context as the United States, no one will notice. Mosquito populations are not monitored in this way. Further attempts could be made if the first one fails. All someone would have to do each time is to go to some sufficiently remote marshy area and release a bunch of mosquitoes. This scenario is not at all far-fetched.

Take another scenario. Though the evidence is not definitive, pollinator declines, particularly of bees and other insects, are believed to be a serious problem for food security in many areas of the world. In the United States, the seriousness of the problem is acknowledged by the Department of Agriculture. We have seen earlier how gene drives could be designed to turn insect vectors of disease sterile. Now, imagine a hostile state agency designing one to spread through pollinator populations. Once again, unnoticed multiple and repeated release of engineered insects would not present a problem; they would almost certainly go unnoticed. Even if the drive does not succeed in spreading through the entire species, it could decrease populations enough to affect agricultural output in states such as California and Florida which have a large variety of pollinator-dependent food crops. Even if the drive lasts a few seasons, the economic and social harm could be huge. This is also not a far-fetched scenario.

Once again, the lesson to be drawn will sound like a broken record. What we need is an ongoing systematic public discussion of our vulnerabilities and a commitment to address them. Very little is being done at present. As we have seen in this book, the potential benefits of CRISPR technology are enormous and we must encourage further research and also ensure that the benefits are distributed in a just fashion. But the same power that makes CRISPR technology the source of so much hope also makes it potentially dangerous, whether it be in the form of immoral genetic enhancement or malicious gene drives. We need dedicated and ongoing public engagement; as with other aspects of CRISPR, so far public engagement has at best been desultory.

CRISPR WITHOUT THE HYPE

"Atoms for peace" was the somewhat odd title of an article published in an issue of *Ladies' Home Journal* in 1955. It predicted that nuclear energy would soon create a world "in which there is no disease[,] . . . in which hunger is unknown[,] . . . where food never rots and crops never spoil . . . and routine household tasks are just a matter of pushing a few buttons[,] . . . a world where no one stokes a furnace or curses the smog."

Sounds like the article from *Wired* with which this book began? Perhaps it does but these remarks were not hype created by yellow journalism. Rather, these were the words of Harold E. Stassen, US President Dwight Eisenhower's Special Assistant on Disarmament. Partly, no doubt, the rhetoric was designed to help make the construction of nuclear reactors more palatable to a suspicious public that had become aware of the horrors unleashed on Hiroshima and Nagasaki. But that is not the whole story. The rhetoric also captured the dizzying dreams of technocrats who had become completely seduced by the promise of nuclear power. If energy became cheap and plentiful, who knew what resources would then be released for the pursuit of other social priorities.

As we now know, it did not work out that way. Some of the developments that Stassen promised indeed came about at least to a limited extent. But that had nothing to do with nuclear power. Food spoils much less easily now than in the early 1950s but that is because of advances in genetics since the 1960s including recombinant DNA techniques and, now, CRISPR-based gene editing. Some household tasks can indeed be automated but that has to do with advances in electronics technology, not nuclear power. Meanwhile, within one generation, nuclear energy became controversial as a few spectacular accidents—Three Mile Island, Chernobyl and, more recently, Fukushima—convinced many societies that risks associated with nuclear plants were intolerable compared to the benefits that they provided. No one has yet satisfactorily solved the problem of disposing radioactive waste from nuclear reactors. Today, even in the face of potentially catastrophic climate change, many countries which have both viable choices continue to opt for fossil fuel over nuclear energy.

Nuclear energy did not create a world without disease; neither will CRISPR even if it leads to the elimination of a suite of severe genetic diseases. Gene drives may well bring the most recalcitrant insect-borne diseases under control. Nuclear energy did not contribute much to ending hunger, if it contributed anything at all; however, the other technologies mentioned earlier, particularly recombinant DNA, have done much to alleviate hunger over the last half-century even as the Earth's population has continued to grow. CRISPR-based techniques will likely contribute even more as we saw at the end of the fourth chapter. This is the hope of CRISPR.

The rest is hype. We will not have designer babies whether or not we decide that it is desirable to do so. Nor will we have taken over all of human evolution: almost certainly, the struggle against evolving and emerging disease will continue to drive human evolution in spite of CRISPR. But our most important struggle will be to ensure that the benefits of CRISPR are shared by all of us and do not only serve the advantaged. We must also ensure that CRISPR technology does not come to be controlled and managed for private benefit by corporate biotech interests including those who have already rushed to claim patents—and that may turn out to be more difficult than all the technical problems faced by CRISPR.

Notes

PREFACE

p. ix: **For the story of the *Wired* article**, see Maben (2016).
p. ix: **Quotes from the *Wired* article** are from Maxmen (2015).
p. ix: **Twitter quotes** are from Maben (2016).
p. x: **History of eugenics**: see Kevles (1985).
p. x: **Involuntary sterilization in California**: Johnson (2013).
p. xii: **Critics of the HGP**: see, e.g., Tauber and Sarkar (1992, 1993).
p. xii: ***Scientific American* article**: Hall (2010).
p. xiii: **US National Academies**: see National Academies of Sciences, Engineering, and Medicine (2017, Chapter 4).
p. xvi: **New report to break the mold**: NAS (2020).
p. xvi: ***Biotech Juggernaut***: Stevens and Newman (2019)
p. xvii: **CRISPR market valuation**: see Anonymous (2018).

CHAPTER 1

p. 1: ***Davenport's Dream***, see Witkowski and Inglis (2008).
p. 2: **Personal information on Davenport** is from Kevles (1984).
pp. 4–5: **Quote** from Davenport (1911), pp. 267–268.
p. 6: **History of IQ testing** is from Kevles (1985), p. 81.
p. 6: **History of immigration policy**, see Reilly (1987), p. 154.
p. 7: **Gentlemen's Agreement with Japan**, see Kevles (1985), p. 96.
p. 7: **History of immigration legislation**, see Kevles (1985), pp. 97–103.
p. 7: **History of sterilization**, see Reilly (1987), p. 154
p. 7: For **Sharp** (and **vasectomy**), see Gugliotta (1998).
p. 7: For **sterilization attempts of the 1920s**, see Reilly (1987), pp. 156–158.

p. 7: For the **model sterilization law**, see Reilly (2015).
p. 8: For the case of **Carry Buck** (including all quotes), see Kevles (1985), pp. 110–112.
p. 8: **History of sterilization**: see Largent (2011).
p. 9: **Continued involuntary sterilization in California**, see Johnson (2013).
p. 9: **Biologists' rejection of involuntary sterilization**, see Reilly (1987), p. 164.
pp. 10–11: **Haldane quotes** are from Haldane (1938b), pp. 86, 96.
p. 11: On **Haldane's influence**, see Reilly (1987), p. 164.
p. 11: For **eugenics in Canada**, see Reilly (2015), p. 337.
p. 11: For **Hitler in prison**, see Reilly (2015).
p. 12: For **Hitler and US eugenics**, see Black (2003), p. 259.
p. 12: For **German sterilization statistics**, see Reilly (2015), p. 358.
p. 13: **Watson's paper** is Watson (2008).
p. 14: **Quotes from that paper** are form Watson (2008), pp. 21, 30, 17, and 11, in order
p. 14: ***Not in Our Genes***, see Lewontin et al. (1984).
pp. 14–15: **Watson's quotes** are from Harmon (2019).
p. 15: **First map of human genes**, see Haldane (1936).
p. 15: **Linkage of hemophilia and color blindness**, see Bell and Haldane (1937).
p. 15: "**Blood Royal**," see Haldane (1938a).
p. 16: For the **quote about Queen Victoria's mutant father**, see Haldane (1938a), p. 135.
p. 16: **Attendance at Third International Congress of Eugenics**, see Kevles (1985, p. 169).
p. 17: **Colchester survey**, see Kevles (1985).
p. 17: **Penrose quote**, see Penrose (1946), p. 951.
p. 17: **Reprinting of *Heredity in Relation to Eugenics***, see Witkowski and Inglis (2008), pp. vii. vii.
p. 17: **Ridley quote** is from Ridley (2008), p. ix.
p. 18: **Genetic screening in the context of the HGP**, see Holtzman (1989).
p. 20: **Psychiatry and genetics**: see Weinberger and Goldman (2008), p. 142.

CHAPTER 2

p. 26: **Quotes**, see Crick (1958, p. 143).
p. 26: **Challenges to biological information**, see Sarkar (1996).
p. 27: **Wholes and parts**, see Sarkar (1998), Chapter 6.
p. 27: **Operons**, see Monod and Jacob (1961).
p. 27: **Allostery**, see Monod et al. (1963).

p. 28: For **Garrod's discovery**, see Garrod (1902).
pp. 28–30: Details of **Pauling's work on sickle cell hemoglobin** are from Strasser (1999).
p. 29: On **Ingram's work**, see Ingram (1956).
p. 29: **Pauling quote** is from Strasser (1999).
p. 30: **Gene editing**, see National Academies of Sciences, Engineering, and Medicine (2017).
p. 30: For details on **PKU**, see Paul and Brosco (2013).
p. 31: **Quotes** are from Rheinberger (2000), pp. 23–24.
p. 32: On **Szylbaski**, see Wirth et al. (2013).
p. 32: For the **Tatum and Ledergerg quotes**, see Wolff and Lederberg (1994), pp. 5, 6.
p. 33: **Sinsheimer quote** is from Sinsheimer (1969), p. 141.
p. 33: On **Davis' defense of genetic engineering**, see Davis (1970), p. 1280.
p. 34: Facts about **Rogers' gene therapy** attempt are from Friedmann (2001)
p. 35: For the **Cline experiment**, see Beutler (2001) and Wirth et al. (2013).
p. 36: For the **Anderson experiment**, see Wirth et al. (2013) and Doudna and Sternberg (2017).
p. 36: For the **Gelsinger case**, see Stolberg (1999) and Wirth et al. (2013).
p. 36: For **problems with leukemia gene therapy**, see Doudna and Sternberg (2017), pp. 20–21.
pp. 36–37: For **gene therapy and cancer**, see Wirth et al. (2013).
p. 37: For **approvals of gene therapy treatments**, see Daley (2020).
p. 37: The **"BC" acronym** is from Urnov (2018).
p. 37: Gene editing information is from Doudna and Sternberg (2017), p. 22.
p. 37: For **Capecchi's work**, see Folger et al. (1982).
p. 37: **Smithies' experiments**, see Smithies et al. (1985).
p. 38: **Recombination model**, see Szostak et al. (1983).
p. 38: For **Jasin's experiments**, see Rouet et al. (1994).
p. 39: **Quote** is from Doudna and Steinberg (2017), p. 30.
p. 39: **Discussion of ZFNs and TALENs** is from Doudna and Sternberg (2017), pp. 31–32, and Chandrasegaran and Carroll (2015).
p. 40: **Carroll quote** is from Chandrasegaran and Carroll (2015), pp. 979–980.

CHAPTER 3

pp. 41–42: **Quotes from Clinton and Blair** are from http://transcripts.cnn.com/TRANSCRIPTS/0006/26/bn.01.html (last accessed October 04, 2018).

p. 42: **Controversy about HGP**, see Tauber and Sarkar (1992).
p. 42: **Talk in 1992**, see Sarkar (1992).
p. 42: For **cost estimates of HGP**, see Bayley (2006).
p. 42: **Transformation of medicine**, see Bodmer and McKie (1994, p. vii).
p. 43: **Gaps in the draft human genome sequence** in 2000, Quackenbush (2011).
p. 43: **Gaps in the draft human genome sequence** in 2010, Angrist (2011), p. 2n.
p. 43: **Gaps in the draft human genome sequence** in 2018, Scheneider et al. (2017).
pp. 43–44: **Collins quotes** are from Collins (1999), pp. 33–35.
pp. 44–45: **CD-CV, GWAS, and SNP**, see Hall (2010).
p. 45: **Genomic variation for diabetes**, see Hall (2010), p. 64.
p. 45: **Controversy over CD-CV**, see Schork et al. (2009).
p. 46: **Evidence for CD-RV**, see McClelland and King (2010).
p. 46: **Weiss quote** is from Hall (2010), p. 62)
pp. 46–47: **Quote** is from Hall (2010). p. 62.
p. 47: **Screening of newborns**, see Texas Department of State Health Services (2018).
p. 48: **Quote** is from Cook-Deegan (1991), p. 594.
p. 48: ***Splicing Life***, see President's Commission for the Study of Ethical Problems in Medicine and Biomedical and Behavioral Research (1982).
p. 48: **European Parliament quoted** from Cook-Deegan (1991), p. 594.
pp. 48–49: **Quote** is from Cook-Deegan (1991), p. 595.
p. 50: **Training of genetic counselors**, see Holtzman (1989) and Tauber and Sarkar (1992).
p. 51: **Asymptomatically ill biological underclass**, see Nelkin and Tancredi (1994).
p. 51: **Genetic discrimination examples** are from Stein (2008) and Angrist (2011), p. 3.
p. 52: **Genetic Information Nondiscrimination Act**, see Angrist (2011, p. 4).
p. 53: **Junk DNA**, see Ohno (1972).
p. 54: **Defense of the HGP**, see Gilbert (1992).
p. 54: **Critique of *the* human genome**, see Sarkar and Tauber (1991) and Tauber and Sarkar (1992).
p. 55: **Provenance of the reference genome**, see Sherman et al. (2018).
p. 55: **Variation from the reference genome**, see Huddleston et al. (2017).
p. 55: **Strangeness of the reference genome**, see Sherman et al. (2018).
p. 55: **Missing base pairs in the reference genome**, see Levy-Sakin et al. (2019).
p. 56: For **genome sizes**, see Sarkar (2015).

p. 56: ***Paris japonica***, see Pellicer et al. (2010).
p. 56: ***Encephalitozoon intestinalis***, see Corradi et al. (2010).
p. 56: For **estimated size of the human genome before 2000**, see Fields et al. (1994).
p. 56: For **estimated size of the human genome after 2000**, see Sarkar (2015).
p. 56: **G-value paradox**, see Hahn and Wray (2002).
p. 57: **Estimated alternative splicing rates**, see Lynch (2007).
pp. 47–48: **Adaptationism**, see Sarkar (2015).
pp. 47–48: **Lynch's model**, see Lynch (2007).
p. 58: **Future of biology** quote, see Collins (2001), p. 77.

CHAPTER 4

p. 60: **CRISPR discovery**, see Ishino et al. (1987); **palindrome and quote** is from p. 5422.
p. 60: **CRISPR in Shigella and** Salmonella, see Nakata et al. (1989).
p. 60: **CRISPR in** Mycobaterium, see Hermans et al. (1991) and Groenen et al. (1993).
p. 60: **CRISPR in** archaea, see Mojica et al. (1993, 1995).
p. 60: For **details on Mojica's** training, see Lander (2016).
pp. 60–61: **Review of CRISPR across taxa**, see Mojica et al. (2000).
p. 61: **Hybridization studies**, see Mojica and Rodriguez (2016).
p. 61: **Distribution of CRISPR in archaea and bacteria as known today,** Han and She (2017) and Ishino et al. (2018).
p. 61: **Proposed names for the CRISPR** structure, see Barrangou and Horvath (2017).
p. 61: **Janssen proposes "CRISPR"**, see Jansen et al. (2002).
p. 61: **Structure of CRISPR**, see Ishino et al. (2018) and Koonin (2019); details are from this source.
p. 62: **Three groups decode function of CRISPR**, see Morange (2015a).
p. 62: **Alicante group**, see Mojica et al. (2005).
p. 62: **History of decoding CRISPR** function, see Lander (2016).
p. 63: **CRISPR absorbing foreign DNA**, see Pourcel et al. (2015).
p. 63: **Third paper on CRISPR function**, see Bolotin et al. (2005).
p. 63: **Koonin's** work, see Makarova et al. (2005).
p. 63: For **Danisco's involvement**, see Lander (2016).
p. 63: **Danisco's previous use of CRISPR**, see Barrangou and Horvath (2017).
p. 63: **Danisco's critical experiment**, see Barrangou et al. (2006).
p. 64: **CRISPR targets DNA**, see Marraffini and Sontheimer (2008).
p. 64: **CRISPR sequence transcription**, see Brouns et al. (2008).
p. 64: **Protospacers**, see Deveaux et al. (2008).
p. 64: **PAM**, see Mojica et al. (2009).

p. 64: **Complementarity of spacer sequence and** target. see Garneau et al. (2010).
p. 64: **Mechanism of CRISPR-based** immunity, see Koonin (2019).
p. 67: **Speed of CRISPR response to viral infection**, see Hynes et al. (2014).
p. 67: **Spacer acquisition and defective phages**, see Hynes et al. (2014).
p. 67: **Spacer acquisition and viral replication**, see Levy et al. (2015).
p. 68: **Charpentier's involvement**, see Deltcheva et al. (2011).
p. 68: **Transfer of CRISPR between** species, see Sapranauskas et al. (2011).
p. 69: **Importance of Makarova et al. review paper**, see Morange (2015b).
p. 69: **Review**, see Marakova et al. (2011a).
p. 69: **Establishment of CRISPR as a gene editing system**, see Jinek et al. (2012).
p. 70: **Quote** from Jinek et al. (2012, p. 820).
p. 70: **Editing eukaryotic cells**, see Cong et al. (2013).
p. 70: **Editing human cells**, see Mali et al. (2012).
p. 71: **Problem of off-target mutations, see** Knott and Doudna (2018).
p. 71: **Delivery problem**, see Glass et al. (2018).
p. 72: **Human point mutations**, see Tang et al. (2017).
p. 72: **Large size of Cas9**, see Glass et al. (2018).
p. 72: **Cas12a**, see Knott and Doudna (2018).
p. 72: **Review of delivery techniques**, see Glass et al. (2018).
p. 73: **Regulatory approval of adenovirus use**, see Wang et al. (2019).
p. 73: **Adenovirus as the most common virus**, see Glass et al. (2019).
p. 73: **Gold nanoparticles**, see Lee et al. (2017).
p. 73: **Addgene**, see Brandt and Barrangou (2019).
pp. 73–75: **CRISPR examples** are from Brandt and Barrangou (2019).
p. 74: **Bushy tomato plants**, see Kwon et al. (2020).

CHAPTER 5

p. *77*: **In-text quote** is from Paul (1998), p. 95.
p. *78*: **Biographical material on Galton** is from Kevles (1985).
p. *78*: **Protestern quote**, see Langkjær-Bain (2019).
p. *79*: **UCL's apology**, see University College London (2021).
p. *80*: **Fisher window**, see Busby (2020).
p. *81*: **Quote** from UNESCO (1952).
p. 82: **First Crick quote** is from Wolstenhome (1962), pp. 64–65.
p. 82: **Second Crick quote** is from Koop (1989).
p. 82: **Littleness of minds**, "A foolish consistency is the hobgoblin of little minds"—Ralph Waldo Emerson.
p. 83: **Eugenics and birth control**, see Paul (1998), pp. 142–143.

p. 84: **Decrease of incidence of Tay-Sachs disease**, see Kaback (2000).
p. 84: For **Singapore**, see Chan (1985); **quote** from p. 708.
p. 85: **Kevles' phrase**, see Kevles (1985).
p. 85: **Saleeby's quote**, see Saleeby (1909), p. 172.
pp. 85–86: **Haldane's arguments**, see Haldane (1938b).
p. 86: **Moderate eugenics**, see Gyngell ad Selgelid (2016).
p. 87: For **Fisher's mating strategy**, see Box (1978).
p. 88: For **hereditarianiam the United States**, see Haller (1963) and Ludmerer (1972).
p. 89: On the **Lysenko affair**, see Joravsky (1970) and Soyfer (1994).
p. 89: On **playing God**, see Cook-Deegan (1991.
p. 90: For **Kitcher on eugenics**, see Kitcher (1996); **quote** from p. 193.
p. 91: For **Kitcher on choosing not to intervene**, see Kitcher (1996), pp. 196–197.
p. 93: For **environmental issues**, see Sarkar (2012).
p. 94: For **nuclear families**, see Murdock (1949).
p. 94: For **extended families**, see Bengston (2001).

CHAPTER 6

pp. 95–97: **Biographical material on Woody Guthrie** is from Arévalo et al. (2001), Ringman (2007), and Santelli (2012).
pp. 97–101: **Details of the He Jiankui affair** are from Greely (2019).
p. 97: **Hong Kong announcement of human germline editing**, see Cohen (2018).
p. 97: **Guangshou of human embryo gene editing**, see Cyranoski (2019a).
p. 97: **CRISPR-based human genome editing**, see Cyraniski (2006).
p. 98: **Academies' statement**, see Dzau et al. (2018).
p. 99: **Academies' 2017 report**, see National Academies of Sciences, Engineering, and Medicine (2017).
p. 99: ***Nature* commentary**, see Lander et al. (2019).
p. 99: **Academies' statement with Royal Society**, see Dzau et al. (2019).
p. 99: **Quote from Chinese authorities**, see Normille (2019), p. 328.
p. 100: For **Chinese regulations**, see Cyranoski (2019b).
p. 100: For **Rebikov**, see Cyranoski (2019c).
p. 100: For **Oviedo convention** and **US regulations**, see Coller (2019).
p. 100: For the **2020 NAS report**, see National Academy of Sciences (2020).
p. 101: For **ethics of human germline editing**, see Coller (2019).
p. 101: For **religious objections to human germline editing**, see Coller (2019).
p. 103: For **disability advocacy**, see Parens and Asch (2000) and Amundson (2005).

p. 103: For **preference for hereditary deafness**, see Bauman (2005) and Mand et al. (2009).

p. 103: **Quote** is from Klein (1980).

p. 103: For **Ringman's arguments**, see Ringman (2007).

p. 105: For **He's scientific results**, see Cohen (2018).

p. 105: For **CRISPR causing off-target mutations**, see Ledford (2020).

p. 106: For **safety problems with He's experiments**, see Cyranoski (2018).

p. 106: **Pleiotropy of *CCR5* mutation**, see Reardon (2019).

p. 106: **Epistasis of *CCR5* mutation**, see Cyranoski (2018b).

p. 106: ***SLC39A8* gene**, see Lander et al. (2019).

pp. 106–107: **Spectacular cases of pleiotropy**, see Rana and Craymer (2018).

p. 107: **Novels written about them**: *Lenin's Brain* by Tilman Spengler (1993); *The Bison: A Novel about the Scientist Who Defied Stalin* by Dainiel Granin (1989).

pp. 107–111: For the **Vogt story**, see Laubichler and Sarkar (2002) and Gregory (2008).

p. 108: **Vogt reports Lenin's outstanding intellect**, see Richter (1991).

p. 108: ***New York Times* publishes Lenin's testament**, see Eastman (1926).

p. 109: **Quotes** from Vogt (1926), pp. 809, 810–811); translations from Laubichler and Sarkar (2001), p. 82n).

p. 110: **Critics of "genetic" terminology for traits, see** Sarkar (1998) and Laubichler and Sarkar (2001).

p. 110: **Quote** from Timoféeff-Ressovsky and Timoféeff-Ressovsky (1926), p. 150n; original translation.

p. 111: **Timoféeff's rising reputation**, see Laubichler and Sarkar (2001).

p. 111: **Discussions of penetrance and expressivity in the HGP**, see Holtzman (1989) and Sarkar (1998).

p. 111: **Supposedly genetic behavioral traits**, see Sarkar (1998), p. 2.

p. 111: **One adjective-one gene**, see Falk (1991, p. 477).

p. 111: ***Glossary***, see Rieger and Michaelis (1954).

p. 111: **Watson's mutations**, see Wheeler et al. (2008).

p. 114: **Complexity of Huntington' disease**, see Arévalo et al. (2001

CHAPTER 7

p. 118: For **definition of enhancement**, see Juengst and Moseley (2019) and Schermer and Bolt (2011), p. 179.

p. 118: For **an attempted purely scientific account of health**, see Boorse (1977).

p. 119: **ADHD statistics**, see Schermer and Bolt (2011), p. 182.

p. 120: For **discrimination on the basis of skin color** and **data on use of skin lightening creams**, see Khan (2018).

p. 121: **Harris' arguments**, see Harris (2007).
p. 121: **Savulescu's arguments**, see Savulescu (2005), Savulescu et al. (2015), and Gyngell and Savulescu (2017).
p. 121: On **Savulescu's public profile**, see Sparrow (2011).
pp. 121–122: **Savulescu quotes** are from Savulescu (2005), p. 38, 39.
p. 122: **Robertson quote** is from Robertson (1994), p. 16.
p. 123: On **Agar's views**, see Agar (2008).
p. 124: **Well-being as a subjective feeling**, see Sparrow (2011), p. 34.
p. 128: For **genetic obsolescence**, see Sparrow (2019).
p. 128: For **human beings as products**, see Habermas (2003).
p. 129: **Details about Plomin** are from Gillborn (2016); the **Plomin quote** is from p. 366; the Cummings quote is from p. 368.
pp. 129–130: **Plomin's schools** is from Asbury and Pomin (2013), p. 178; quotes are from pp. 187, 3–4, 178–179, in order
p. 129: For **problems with defining traits**, see Sarkar (1998).
p. 131: For **intelligence as a trait**, see Richardson (2000).
p. 131: For **Mugni and Carugati's experiments**, see Mugny and Carugati (1989); **quote** from Richardson (2000), p. 3.
p. 131: **Sternberg's experiments** are described in Richardson (2000), p. 4.
p. 131: **Cultural basis of intelligence**, Mugni and Carugati quoted from Richardson (2000), p. 19.
p. 131: **Sternberg** quoted from Richardson (2000), p. 19.
p. 131: ***Mismeasure of Man***, see Gould (1981).
p. 131: **IQ and racism**, see Gillborn (2016).
p. 131: **Statistics on immigrants**, see Richardson (2000), p. 30.
p. 132: ***Bell Curve***, see Herrnstein and Murray (1994).
p. 132: ***Wall Street Journal***, see Gillborn (2016).
pp. 132–33: **Binet's test and IQ methodology**, see Richardson (2000), pp. 28–36, 42–45.
p. 133: **IQ and Betting**, see Ceci and Liker (1994).
p. 133: **Flynn effect**, see Flynn (1987).
p. 134: **IQ and racism**, see Gillborn (2016).
p. 134: **HGP and IQ**, Sarkar (1998).
p. 134: **New genetics of IQ**, see Plomin and von Stumm (2018).
p. 134: For **heritability**, see Sarkar (1998), Chapter 4.
p. 136: **Heritability of height**, see McEvoy and Visscher (2009).
p. 136: For **indirect estimates of heritability**, see Sarkar (1998).
p. 137: **GWAS and heritability of IQ**, see Plomin and van Stumm (2018).
p. 137: **GPS predicts IQ**, see Plomin and von Stumm (2018), p. 150.
p. 138: **Complexity of GPS**, see Plomin and von Stumm (2018), p. 156, Box 7.
pp. 138–139: **Rosenberg and collaborators on GPS**, see Rosenberg et al. (2019); **quotes** are from p. 27..
p. 140: **Single gene for eye color**, see Davenport and Davenport (1907) and Hurst (1908).

p. 140: For **exceptions to single gene model for eye color**, see Sturm and Larsson (2009).
p. 140: For **complexity of eye color phenotype**, see White and Rabago-Smith (2011).
p. 140: **Two main genes for eye color**, see Jablonski (2018).
p. 140: For **brown-eyed child of blue-eyed parents**, see White and Rabago-Smith (2011), p. 6.
p. 141: **Sixteen genes for eye color**, see Sturm and Larsson (2009).
p. 141: **GWAS for eye color**, see Maroñas et al. (2015).
p. 141: **Melanin composition rules for hair color**, see Maroñas et al. (2015, p. 16).
p. 141: **Green eyes and red hair**, see Barsh (2003).
p. 142: **GWAS for hair color**, see Pavan and Sturm (2019).
p. 142: **Two types of skin pigmentation**, Maroñas et al. (2015).
p. 142: **Barsh quote** is from Barsh (2003), p. 19.
pp. 142–143: **Genes in South Asian and European populations**, see Maroñas et al. (2015), p. 27).
p. 143: **Genes in African populations**, see Pavan and Sturm (2019), pp. 58–59.
p. 143: For **reductionism**, see Sarkar (1998).
p. 144: For **critique of the HGP**, see Tauber and Sarkar (1992, 1993), Sarkar (1998).
p. 146: **Chick and frog embryogenesis**, see Ruden et al. (2018).
p. 146: **Symbionts in animal development**, see Sariola and Gilbert (2020).
p. 146: **Temperature dependence of reptile sex**, see Bull (1983).
p. 146: **Amphipod sex**, see McCabe and Dunn (1997).
p. 146: ***Bonnelia viridis***, see Leutert (1975).
p. 146: On **phenotypic plasticity**, see Sarkar (1999).
p. 146: **European map butterfly**, see van der Weele (1995).
p. 146: ***Daphnia cucullata***, see Agrawal et al. (1999).
p. 146: **Phenotypic plasticity in humans**, see Feinberg (2007, p. 433).
p. 147: ***New York Times* interview**, see Laurence (1932).
p. 147: For **Haldane's changing views on eugenics**, see Haldane (1938b) and Paul (1998).

CHAPTER 8

p. 152: **Selfish DNA used to target disease vectors**, see Burt (2003).
p. 152: **Bier and Gantz's gene drive**, see Gantz and Bier (2015).
p. 153: **Targeting *Anopheles stephensi***, see Gantz et al. (2015).
p. 153: **Targeting *Anopheles gambiae***, see Hammond et al. (2016).
p. 153: For potentially **targeting *Aedes aegupti***, see Hall et al. (2015).
p. 153: For **problems with gene drives**, see Sarkar (2018).

p. 154: **Evolution limits gene drives**, see Bull and Malik (2017).
p. 154: **Mating patterns limits gene drives**, see Shaw (2016).
p. 154: **Resistance alleles for gene drives**, see Champer et al. (2017).
p. 154: **Resistance alleles for in Imperial College experiments**, see Hammond et al. (2017).
p. 154: **Natural variation limits gene drives**, see Zentner and Wade (2017).
p. 155: **Blocking resistance allele mutations**, see Champer et al. (2018).
p. 156: ***P*-element**, see Rode et al. (2019).
p. 156: **Gene drives against rodents**, see Godwin et al. (2019).
p. 156: **Rodents' effects on biodiversity**, see Leitschuh et al. (2018).
p. 156: **Controlling rodents**, see Piaggio et al. (2017).
p. 156: **Gene drives disrupting ecology**, see Backus and Gross (2016).
p. 156: **Vectors for dengue and other flaviviruses**, see Gardner and Sarkar (2013).
p. 156: **2015 Zika crisis**, see Sarkar and Gardner (2016).
p. 157: For **mosquito elimination examples**, see Löwy (2017).
p. 157: For **yellow fever resurgence after DDT ban**, see Moreno-Madriñán and Turell (2018).
p. 157: **Daisy-chain gene drives**, see Noble et al. (2019).
p. 158: For **state actors and CRISPR**, see Ouagrham-Gormley and Fye-Marnien (2019).
p. 159: **Feith quote** is from Vogel and Ouagrham-Gormley (2018), p. 203.
p. 159: **Agencies investigating CRISPR-based threats**, see Bagley (2015).
p. 159: For **undergraduates designing a gene drive**, see Swelitz (2016); the team consisted of Kathryn Almquist, Chase Bowen, Carolyn Domroese, Ajinkya Limkar, Sarah Lucas,and Sophie Vrba.
p. 159: For **DIY Bio movement**, see Gronvall (2019).
p. 160: For **biosecurity experts worrying about CRISPR**, see Ouagrham-Gormley and Fye-Marnien (2019).
p. 160: **Quote from** Vogel and Ouagrham-Gormley (2018), p. 204.
p. 161: For **pollinator problems**, Ramaswamy (2017).
pp. 161–162: **Quote** from Vogel and Ouagrham-Gormley (2018), p. 203.

Acknowledgments

I must begin by thanking Harvey Shoolman for convincing me to use a "cut-and-paste" title instead of a more banal one I originally intended. Lenny Moss read and commented on the entire book besides being a sounding board for my ideas over many years. Stuart Newman has played the same role in recent years. Christopher Donohue, Alan Love, and Frankie Mace provided comments on individual chapters. The material on which this book is based was presented at several institutions including the Illinois Institute of Technology, Michigan State University, the National Human Genome Research Institute (in Bethesda, Maryland), Presidency University (Kolkata), the University of Miami, the University of Rijeka, and the University of Texas at Austin, as well as at venues including the Austin Forum and Health Tech Austin. I thank the organizers of those talks for the invitations and the audiences for feedback.

Over the years, six individuals have influenced my views on genes and what they do. Dick Lewontin influenced me both through his writing and his mentorship during the year that I spent in his laboratory. I learned most of my population genetics from Jim Crow who also taught me to have a more nuanced view of eugenics than was common in the 1980s and 1990s. Rafi Falk made alive the Byzantine world of late classical genetics and, in the process, we wrote several papers together. The critique of the Human Genome Project presented in the third chapter of this book was originally developed in collaboration with Fred Tauber. Scott Gilbert and I have only written one paper together. But we have been bemoaning the lure of genetic reductionism for decades. Lenny Moss has already been mentioned.

I have had many discussions of CRISPR with Zach Adelman, Ilya Finkelstein, Joanna Masel, Stuart Newman, and Arlin Stoltzfus. Adelman also has been a source of insight on gene drives and some of that work has previously been published in *BioScience*. I have discussed eugenics with so many people over the years that listing them would require at least an entire chapter. However, Diane Paul deserves special mention. Some of the work on Vogt and Lenin's brain was done jointly with Manfred Laubichler and originally published with him. It was supported by the Wissehschaftskolleg zu Berlin through a Fellowship during 1996–1997.

Laurenz Casser checked some of the translations of quotations from German. At Rowman & Littlefield, this book was initially sponsored by Isobel Cowper-Coles who provided very pertinent feedback during the early stages of this project. As the project drew to completion, Frankie Mace and Scarlet Furness took over that role and shepherded this book to print.

Finally, my wife, Katherine Dunlop, and my two daughters, Kajri Marilyn Sarkar and Malini Leslie Sarkar, had to bear with me as the book took away my time from them especially in the middle of a pandemic that disorganized all our lives in earnest.

References

Agar, N. 2008. How to defend genetic enhancement. In Gordijn, B., and Chadwick, R., Eds. *Medical Enhancement and Posthumanity*. Dordrecht: Springer, pp. 55–67.

Agrawal, A. A., Laforsch, C., and Tollrian, R. 1999. Transgenerational induction of defences in animals and plants. *Nature* 40: 60–63.

Amundson, R. 2005. Disability, ideology, and quality of life. In Wasserman, D., Bicjenbach, J., and Wachbroit, R., Eds. *Quality of Life and Human Difference: Genetic Testing, Health Care and Disability*. Cambridge, UK: Cambridge University Press, pp. 101–120.

Angrist, M. 2011. *Here is a Human Being: At the Dawn of Personal Genomics*. New York: Harper Collins.

Anonymous. 2018. *Press Release: CRISPR Cas9 Genome Editing Market Worth $5.3 Billion by 2025*. https://markets.businessinsider.com/news/stocks/crispr-cas9-genome-editing-market-worth-5-3-billion-by-2025-1027738971. Last accessed 17 February 2020.

Arévalo, J., Wojcieszek, J., and Conneally, P. M. 2001. Tracing Woody Guthrie and Huntington's disease. *Seminars in Neurology* 21: 209–224.

Asbury, K., and Plomin, R. 2013. *G is for Genes: The Impact of Genetics on Education and Achievement*. New York: John Wiley and Sons.

Backus, G. A., and Gross, K. 2016. Genetic engineering to eradicate invasive mice on islands: Modeling the efficiency and ecological impacts. *Ecosphere* 7: e01589.

Bagley, S. 2015. Why the FBI and Pentagon are afraid of this new genetic technology. *STAT News*. https://www.statnews.com/2015/11/12/gene-drive-bioterror-risk/. Last accessed 12 November 2019.

Baltimore, D., Berg, P., Botchan, M., Carroll, D., Charo, R. A., Church, G., Corn, J. E., Daley, G. Q., Doudna, J. A., Fenner, M., Greely, H. T., Jinek, M., Martin, G. S., Penhoet, E., Puck, J., Sternberg, S. H., Weissman, J. S., and Yamamoto, K. R. 2015. A prudent path forward for genomic engineering and germline gene modification. *Science* 348: 36–38.

Barrangou, R., Fremaux, C., Deveau, H., Richards, M., Boyaval, P., Moineau, S., Romero, D. A., and Horvath, P. 2007. CRISPR provides acquired resistance against viruses in prokaryotes. *Science* 315: 1709–1712.

Barrangou, R., and Horvath, P. 2017. A decade of discovery: CRISPR functions and applications. *Nature Microbiology* 2: 17092.

Barsh, G. S. 2003. What controls variation in human skin color? *PLoS Biology* 1(1): e27. doi: 10.1371/jour- nal.pbio.0000027.

Bauman, H. D. L. 2005. Designing deaf babies and the question of disability. *The Journal of Deaf Studies and Deaf Education* 10: 311–315.

Bayley, H. 2006. Sequencing single molecules of DNA. *Current Opinion in Chemical Biology* 10: 628–637.

Bearn, A. G. 1993. *Archibald Garrod and the Individuality of Man.* Oxford: Clarendon Press.

Bearn, A. G., and Miller, E. D. 1979. Archibald Garrod and the development of the concept of inborn errors of metabolism. *Bulletin of the History of Medicine* 53: 315–328.

Bell, J., and Haldane, J. B. S. 1937. The linkage between the genes for colour-blindness and haemophilia in man. *Proceedings of the Royal Society of London B* 123: 119–150.

Bengston, V. L. 2001. Beyond the nuclear family: The increasing importance of multigenerational bonds. *Journal of Marriage and Family* 63: 1–16.

Beutler, E. 2001. The Cline affair. *Molecular Therapy* 4: 396–397.

Black, E. 2003. *War Against the Weak: Eugenics and America's Campaign to Create a Master Race*. New York: Four Walls Eight Windows.

Bodmer, W., and McKie, R. 1994. *The Book of Man: The Quest to Discover Our Genetic Heritage*. New York: Viking.

Bolotin, A., Quinquis, B., Sorokin, A., and Ehrlich, S. D. 2005. Clustered regularly interspaced short palindrome repeats (CRISPRs) have spacers of extrachromosomal origin. *Microbiology* 151: 2551–2561.

Boorse, C. 1977. Health as a scientific concept. *Philosophy of Science* 44: 542–573.

Box, J. F. 1978. *R. A. Fisher: The Life of a Scientist*. New York: Wiley.

Brandt, K., and Barrangou, R. 2019. Applications of CRISPR technologies across the food supply chain. *Annual Review of Food Science and Technology* 10: 133–150.

Brossard, D., Belluck, P., Gould, F., and Wirz, C. D. 2019. Promises and perils of gene drives: Navigating the communication of complex, post-normal science. *Proceedings of the National Academy of Sciences* 116: 7692–7697.

Brouns, S. J. J., Jore, M. M., Lundgren, M., Westra, E. R., Slijkhuis, R. J. H., Snijders, A. P. L., Dickman, M. J., Makarova, K. S., Koonin, E. V., and van der Oost, J. 2008. Small CRISPR RNAs guide antiviral defense in prokaryotes. *Science* 321: 960–964.

Bull, J. J. 1983. *Evolution of Sex Determining Mechanisms*. Menlo Park, CA: Benjamin/Cummings Publishing Company.

Bull, J. J., and Malik, H. S. 2017. The gene drive bubble: New realities. *PLoS Genetics* 13: e1006850.

Burt, A. 2003. Site-specific selfish genes as tools for the control and genetic engineering of natural populations. *Proceedings of the Royal Society of London B* 270: 921–928.

Busby, M. 2020. Cambridge college to remove window commemorating eugenicist. *The Guardian*, 27 June. https://www.theguardian.com/education/2020/jun/27/cambridge-gonville-caius-college-eugenicist-window-ronald-fisher. Last accessed 13 November 2020.

Ceci, S. J., and Liker, J. K. 1986. A day at the races: A study of IQ, expertise, and cognitive complexity. *Journal of Experimental Psychology* 115: 255–256.

Champer, J., Liu, J., Oh, S. Y., Reeves, R., Luthra, A., Oakes, N., Clark, A. G., and Messer, P. W. 2018. Reducing resistance allele formation in CRISPR gene drive. *Proceedings of the National Academy of Sciences (USA)* 115: 5522–5527.

Champer, J., Reeves, R., Oh, S. Y., Liu, C., Liu, J., Clark, A. G., and Messer, P. W. 2017. Novel CRISPR/Cas9 gene drive constructs reveal insights into mechanisms of resistance allele formation and drive efficiency in genetically diverse populations. *PLoS Genetics* 13: e1006796.

Chan, C. K. 1985. Eugenics on the rise: a report from Singapore. *International Journal of Health Services* 15: 707–712.

Chandrasegaran, S., and Carroll, D. 2016. Origins of programmable nucleases for genome engineering. *Journal of Molecular Biology* 428: 963–989.

Cohen, J. 2018. What now for human genome editing? *Science* 363: 1090–1092.

Cohen, J., and Normille, D. 2020. China delivers verdict on gene editing of babies. *Science* 367: 130.

Coller, B. S. 2019. Ethics of human genome editing. *Annual Review of Medicine* 70: 289–305.

Collins, F. S. 1991. The genome project and human health. *FASEB Journal* 5: 77.

Collins, F. S. 1999. Medical and societal consequences of the human genome project. *New England Journal of Medicine* 341: 28–37.

Collins, F. S., and Varmus, H. 2015. A new initiative on precision medicine. *New England Journal of Medicine* 372: 793–795.

Comfort, N. 2018. Can we cure genetic diseases without slipping into eugenics? In Obasogie, O. K., and Darnovsky, M., Eds. *Beyond Bioethics: Toward a New Biopolitics*. Berkeley: University of California Press, pp. 175–185.

Cong, L., Ran, F. A., Cox, D., Lin, S., Barretto, R., Habib, N., Hsu, P. D., Wu, X., Jiang, W., Marraffini, L. A., and Zhang, F. 2013. Multiplex genome engineering using CRISPR/Cas systems. *Science* 33: 819–823.

Cook-Deegan, R. M. 1991. Mapping the human genome. *Southern California Law Review* 65: 579–596.

Corradi, N., Pombert, J.-F., Farinelli, L., Didier, E. S., and Keeling, P. K. 2010. The complete sequence of the smallest known nuclear genome from the microsporidian *Encephalitozoon intestinalis*. *Nature Communications* 1. doi:10.1038/ncomms1082.

Crick, F. H. C. 1958. On protein synthesis. *Symposium of the Society for Experimental Biology* 12: 138–163.

Cyranoski, D. 2016. First trial of CRISPR in people. *Nature* 535: 476–477.

Cyranoski, D. 2018a. First CRISPR babies: 6 questions that remain. *Nature*, 2 December. https://www.scientificamerican.com/article/first-crispr-babies-6-questions-that-remain/. Last accessed 10 July 2019.

Cyranoski, D. 2018b. Baby gene edits could affect a range of traits. *Nature*, 12 December. https://www. nature.com/articles/d41586-018-07713-2. Last accessed 10 July 2019.

Cyranoski, D. 2019a. The CRISPR-baby scandal: What's next for human gene-editing. *Nature* 566: 440–442.

Cyranoski, D. 2019b. China set to introduce gene-editing regulation following CRISPR-baby furore. *Nature*. doi:10.1038/d41586-019-01580-1.

Cyranoski, D. 2019c. Russian biologist plans more CRISPR-edited babies. *Nature* 570: 145.

Cyranoski, D., and Reardon, S. 2015. Chinese scientists genetically modify human embryos. *Nature News*, 22 April. doi:10.1038/nature.2015.17378. Last accessed 11 October 2018.

Daley, J. 2020. Gene therapy arrives. *Scientific American*. https://www.scientificamerican.com/article/ gene-therapy-arrives/. Last accessed 25 June 2020.

Davenport, C. B. 1911. *Heredity in Relation to Eugenics*. New York: Henry Holt and Company.

Davenport, G. C., and Davenport, C. B. 1907. Heredity of eye-color in man. *Science* 26: 589–592.

Davis, B. D. 1970. Prospects for genetic intervention in man. *Science* 170: 1279–1283.

Deltcheva, E., Chylinski, K., Sharma, C. M., Gonzales, K., Chao, Y., Pirzada, Z. A., Eckert, M. R., Vogel, J., and Charpentier, E. 2011. CRISPR RNA maturation by trans-encoded small RNA and host factor RNase III. *Nature* 471: 602–607.

Deveau, H., Barrangou, R., Garneau, J. E., Labont´e, J., Fremaux, C., Boyaval, P., Romero, D. A., Horvath, P., and Moineau, S. 2008. Phage response to CRISPR-encoded resistance in Streptococcus thermophilus. *Journal of Bacteriology* 190: 1390–1400.

Doudna, J. A., and Sternberg, S. H. 2017. *A Crack in Creation: Gene Editing and the Unthinkable Power to Control Evolution*. Boston: Houghton Mifflin Harcourt.

Dzau, V. J., McNutt, M., and Bai, C. 2018. Wake-up call from Hong Kong. *Science* 362: 1215.

Dzau, V. J., McNutt, M., and Ramakrishnan, V. 2019. Academies' action plan for germline editing. *Nature* 567: 175.

Eastman, M. 1926. Lenin's 'testament' at last revealed: Letter, hidden after leader's death, warned against Stalin and extolled Trotsky, 18 October, p. 1.

Falk, R. 1991. The dominance of traits in genetic analysis. *Journal of the History of Biology* 24: 457–484.

Feinberg, A. 2007. Phenotypic plasticity and the epigenetics of human disease. *Nature* 447: 433–440.

Fields, C., Adams, M. D., White, O., and Venter, J. C. 1994. How many genes in the human genome? *Nature Genetics* 7: 345–346.

Flynn, J. R. 1987. Massive IQ gains in 14 nations: What IQ tests really measure. *Psychological Bulletin* 101: 171–191.

Folger, K. R., Wong, E. A., Wahl, G., and Capecchi, M. R. 1982. Patterns of integration of DNA microinjected into cultured mammalian cells: Evidence for homologous recombination between injected plasmid DNA molecules. *Molecular and Cellular Biology* 2: 1372–1387.

Gantz, V. M., and Bier, E. 2015. The mutagenic chain reaction: A method for converting heterozygous to homozygous mutations. *Science* 348: 442–444.

Gantz, V. M., Jasinskiene, N., Tatarenkova, O., Fazekas, A., Macias, V. M., Bier, E., and James, A. A. 2015. Highly efficient Cas9-mediated gene drive for population modification of the malaria vector mosquito *Anopheles stephensi*. *Proceedings of the National Academy of Sciences (USA)* 112: E6736–E6743.

Gardner, L., and Sarkar, S. 2013. A global airport-based risk model for the spread of dengue infection via the air transport network. *PLoS One* 8: e72129.

Garneau, J .E., Dupuis, M.-É., Villion, M., Romero, D. A., Barrangou, R., Boyaval, P., Fremaux, C., Horvath, P., Magadán, A. H., and Moineau, S. 2010. The CRISPR/Cas bacterial immune system cleaves bacteriophage and plasmid DNA. *Nature* 468: 67–71.

Garrod, A. 1902. The incidence of alkaptonuria: A study in chemical individuality. *Lancet* 160: 1616–1620.

Gilbert, W. 1992. A vision of the grail. In Kevles, D. J., and Hood, L., Eds. *The Code of Codes: Scientific and Social Issues in the Human Genome Project*. Cambridge, MA: Harvard University Press, pp. 83–97.

Gillborn, D. 2016. Softly, softly: Genetics, intelligence and the hidden racism of the new geneism. *Journal of Education Policy* 31: 365–388.

Glass, Z., Lee, M., Li, Y., and Xu, Q. 2018. Engineering the delivery system for CRISPR-based genome editing. *Trends in Biotechnology* 36: 173–185.

Godwin, J., Serr, M., Barnhill-Dilling, S. K., Blondel, D. V., Brown, P. R., Campbell, K., Delborne, J., Lloyd, A. L., Oh, K. P., Prowse, T. A., Saah, R., and Thomas, P. 2019. Rodent gene drives for conservation: Opportunities and data needs. *Proceedings of the Royal Society B* 286: 20191606.

Gould, S. J. 1981. *Mismeasure of Man*. New York: W. W. Norton and Sons.

Granin, D. 1989. *The Bison: A Novel About the Scientist Who Defied Stalin*. New York: Doubleday.

Greely, H. T. 2019. CRISPR'd babies: Human germline genome editing in the 'He Jiankui affair.' *Journal of Law and the Biosciences* 6: 111–183.

Gregory, P. R. 2008. *Lenin's Brain and Other Tales from the Secret Soviet Archives*. Stanford, CA: Hoover Institution Press.

Groenen, P. M., Bunschoten, A. E., van Soolingen, D., and van Embden, J. D. 1993. Nature of DNA polymorphism in the direct repeat cluster of *Mycobacterium tuberculosis*: Application for strain differentiation by a novel typing method. *Molecular Microbiology* 10: 1057–1065.

Gronvall, G. K. 2019. Synthetic biology: Biosecurity and biosafety implications. In Singh, S. K., and Kuhn, J. H., Eds. *Defense Against Biological Attacks*. Cham: Springer, pp. 225–232.

Gugliotta, A. 1998. "Dr. Sharp with his little knife": Therapeutic and punitive origins of eugenic vasectomy—Indiana, 1892–1921. *Journal of the History of Medicine and Allied Sciences* 53: 371–406.

Guttinger, S. 2018. Trust in science: CRISPR–Cas9 and the ban on human germline editing. *Science and Engineering Ethics* 24: 1077–1096.

Gyngell, C., and Savulescu, J. 2018. The simple case for germline gene editing. In Delatycki, M., Blashki, G., and Sykes, H., Eds. *Genes for Life: The Impact of the Genetic Revolution.* Albert Park, Victoria: Future Leaders, pp. 28–45.

Gyngell, C., and Selgelid, M. 2016. Twenty-first century eugenics. In Francis, L., Ed. *The Oxford Handbook of Reproductive Ethics.* New York: Oxford University Press, pp. 141–158.

Habermas, J. 2003. *The Future of Human Nature.* Cambridge, MA: Polity Press.

Hahn, M. W., and Wray, G. A. 2002. The G-value paradox. *Evolution and Development* 4: 73–75.

Haldane, J. B. S. 1936. A provisional map of a human chromosome. *Nature* 137: 398–400.

Haldane, J. B. S. 1938a. Blood royal: A study of hemophilia in the royal families of Europe. *Modern Quarterly* 1: 129–139.

Haldane, J. B. S. 1938b. *Heredity and Politics.* New York: W. W. Norton.

Hall, A. B., Basu, S., Jiang, X., Qi, Y., Timoshevskiy, V. A., Biedler, J. K., Sharakhova, M. V., Elahi, R., Anderson, M. A., Chen, X. G., Sharakhov, I. V., and Adelman, Z. N. 2015. A male-determining factor in the mosquito *Aedes aegypti. Science* 348: 1268–1270.

Hall, S. S. 2010. Revolution postponed. *Scientific American* 303(4): 60–67.

Haller, M. H. 1963. *Eugenics: Hereditarian Attitudes in American Thought.* New Brunswick: Rutgers University Press.

Hammond, A., Galizi, R., Kyrou, K., Simoni, A., Siniscalchi, C., Katsanos, D., Gribble, M., Baker, D., Marois, E., Russell, S., Burt, A., Windbichler, W., Crisanti, A., and Nolan, T. 2016. A CRISPR-Cas9 gene drive system targeting female reproduction in the malaria mosquito vector *Anopheles gambiae. Nature Biotechnology* 34: 78–83.

Hammond, A. M., Kyrou, K., Bruttini, M., North, A., Galizi, R., Karlsson, X., Kranjc, N., Carpi, F. M., D'Aurizio, R., Crisanti, A., and Nolan, T. 2017. The creation and selection of mutations resistant to a gene drive over multiple generations in the malaria mosquito. *PLoS Genetics* 13: e1007039.

Han, W., and She, Q. 2017. CRISPR history: Discovery, characterization, and prosperity. *Progress in Molecular Biology and Translational Science* 152: 1–21.

Harmon, A. 2019. James Watson had a chance to salvage his reputation on race. He made things worse. *New York Times*, 1 January. https://www.nytimes.com/2019/01/01/science/watson-dna-genetics-race. html. Last accessed 2 March 2019.

Harris, J. 2010. *Enhancing Evolution: The Ethical Case for Making Better People.* Princeton: Princeton University Press.

Hermans, P. W., van Soolingen, D., Bik, E. M., de Haas, P. E., Dale, J. W., and van Embden, J. D. 1991. Insertion element IS987 from *Mycobacterium bovis* BCG is located in a hot-spot integration region for insertion elements in *Mycobacterium tuberculosis* complex strains. *Infection and Immunity* 59: 2695–2705.

Herrnstein, R. J., and Murray, C. 1994. *The Bell Curve: Intelligence and Class Structure in American Life*. New York: Free Press.

Holtzman, N. A. 1989. *Proceed with Caution: Predicting Genetic Risks in the Recombinant DNA Era*. Baltimore: Johns Hopkins University Press.

Huddleston, J., Chaisson, M. J., Steinberg, K. M., Warren, W., Hoekzema, K., Gordon, D., Graves-Lindsay, T. A., Munson, K. M., Kronenberg, Z. N., Vives, L., Peluso, P., Boitano, M., Chin, C.-S., Korlach, J., Wilson, R. K., and Eichler, E. E. 2017. Discovery and genotyping of structural variation from long-read haploid genome sequence data. *Genome Research* 27: 677–685.

Hurst, C. C. 1908. On the inheritance of eye-colour in man. *Proceedings of the Royal Society of London B* 80: 85–96.

Hynes, A. P., Villion, M., and Moineau, S. 2014. Adaptation in bacterial CRISPR-Cas immunity can be driven by defective phages. *Nature Communications* 5: 4399.

Ingram, V. M. 1956. A specific chemical difference between globins of normal and sickle-cell Anemia hemoglobins. *Nature* 178: 792–794.

Ishino, Y., Krupovic, M., and Forterre, P. 2018. History of CRISPR-Cas from encounter with a mysterious repeated sequence to genome editing technology. *Journal of Bacteriology* 200: e00580-17.

Ishino, Y., Shinagawa, H., Makino, K., Amemura, M., and Nakata, A. 1987. Nucleotide sequence of the *iap* gene, responsible for alkaline phosphatase isozyme conversion in *Escherichia coli*, and identification of the gene product. *Journal of Bacteriology* 169: 5429–5433.

Jablonski, M. 2018. Eye color. In Trevathan, W., Ed. *International Encyclopedia of Biological Anthropology*. New York: John Wiley & Sons, pp. 1–2.

Jansen, R., Embden, J. D., Gaastra, W., and Schouls, L. W. 2002. Identification of genes that are associated with DNA repeats in prokaryotes. *Molecular Microbiology* 43: 1565–1575.

Janssens, A. C. J. W. 2016. Designing babies through gene editing: Science or science fiction? *Genetics in Medicine* 18: 1186–1187.

Jinek, M., Chylinski, K., Fonfara, I., Hauer, M., Doudna, J. A., and Charpentier, E. 2012. A programmable dual-RNA-guided DNA endonuclease in adaptive bacterial immunity. *Science* 337: 816–821.

Johnson, C. G. 2013. Female inmates sterilized in California prisons without approval. *NBC Bay Area*. https://www.nbcbayarea.com/news/california/Female-Inmates-Sterilized-in-California-Prisons-Without-Approval-214634341.html. Last accessed 19 August 2019.

Joravsky, D. 1970. *Lysenko Affair*. Chicago: University of Chicago Press.

Juengst, E., and Moseley, D. 2019. Human enhancement. In Zalta, E. N., Ed. *Stanford Encyclopedia of Philosophy*. https://plato.stanford.edu/archives/sum2019/entries/enhancement/. Last accessed 12 November 2019.

Kaback, M. M. 2000. Population-based genetic screening for reproductive counseling: The Tay-Sachs disease model. *European Journal of Pediatrics* 159: S192–S195.

Kevles, D. J. 1985. *In the Name of Eugenics: Genetics and the Uses of Human Heredity*. Berkeley: University of California Press.

Khan, C. 2018.Skin-lightening creams are dangerous—Yet business is booming. Can the trade be stopped? *Guardian*, 23 April. https://www.theguardian.com/world/2018/apr/23/skin-lightening-creams-are-dangerous-yet-business-is-booming-can-the-trade-be-stopped. Last accessed 12 December 2019.

Kitcher, P. 1996. *The Lives to Come: The Genetic Revolution and Human Possibilities*. New York: Simon and Schuster.

Klein, J. 1980. *Woody Guthrie: A Life*. New York: Alfred Knopf.

Knott, G. J., and Doudna, J. A. 2018. CRISPR-Cas guides the future of genetic engineering. *Science* 361: 866–869.

Koonin, E. V. 2019. CRISPR: A new principle of genome engineering linked to conceptual shifts in evolutionary biology. *Biology and Philosophy* 34: 9.

Koop, C. (1989). Life and death and the handicapped newborn. *Issues in Law & Medicine*, 5: 101–116.

Kozubek, J. 2018. *Modern Prometheus: Editing the Human Genome with Crispr-Cas9*. New York: Cambridge University Press.

Krimsky, S. 1982. *Genetic Alchemy*. Cambridge, MA: MIT Press.

Kwon, C. T., Heo, J., Lemmon, Z. H., Capua, Y., Hutton, S. F., Van Eck, J., Park, S. J., and Lippman, Z. B. 2020. Rapid customization of Solanaceae fruit crops for urban agriculture. *Nature Biotechnology* 38: 182–188.

Lander, E. S. 2016. Heroes of CRISPR. *Cell* 164: 18–28.

Lander, E. S., Baylis, F., Zhang, F., Charpentier, E., Berg, P., Bourgain, C., Friedrich, B., Joung, J. K., Li, J., Liu, D, Naldini, L., Nie, J. B., Qiu, R., Schoene-Seifert, B., Shao, F., Terry, S., Wei, W., and Winnacker, E. L. 2019. Adopt a moratorium on heritable genome editing. *Nature* 567: 165–168.

Langkjær-Bain, R. 2019. The troubling legacy of Francis Galton. *Significance* 16(3): 16–21.

Lanphier, E., Urnov, F., Haecker, S. E., Werner, M., and Smolenski, J. 2015. Don't edit the human germ line. *Nature* 519: 410–411.

Largent, M. A. 2011. *Breeding Contempt: The History of Coerced Sterilization in the United States*. New Brunswick, NJ: Rutgers University Press.

Laubichler, M., and Sarkar, S. 2002. Flies, genes, and brains: Oskar Vogt, Nikolai Timoféeff-Ressovsky, and the origin of the concepts of penetrance and expressivity. In Parker, L. S., and Ankeny, R., Eds. *Mutating Concepts, Evolving Disciplines: Genetics, Medicine, and Society*. Dordrecht: Kluwer, pp. 63–85.

Laurence, W. E. 1932. Not a "perfect man" in Haldane's Utopia. *New York Times*, 29 August, p. 15.

Ledford, H. 2020. CRISPR editing wreaks chromosomal mayhem in human embryos. *Nature* 583: 17–18.

Lee, K., Conboy, M., Park, H. M., Jiang, F., Kim, H. J., Dewitt, M. A., Mackley, V. A., Chang, K., Rao, A., Skinner, C., Shobha, T., Mehdipour, M., Liu, H., Huang, W.-C., Lan, F., Bray, N., Li, S., Corn, J. E., Kataoka, K., Doudna, J. A., Conboy, I., and Murthy, N. 2017. Nanoparticle delivery of Cas9 ribonucleoprotein and donor DNA *in viv* o induces homology-directed DNA repair. *Nature Biomedical Engineering* 1: 889–901.

Leitschuh, C. M., Kanavy, D., Backus, G. A., Valdez, R. X., Serr, M., Pitts, E. A., Threadgill, D., and Godwin, J. 2018. Developing gene drive technologies to eradicate invasive rodents from islands. *Journal of Responsible Innovation* 5: S121–S138.

Leutert, R. 1975. Sex-determination in *Bonellia*. In Reinboth, R., Ed. *Intersexuality in the Animal Kingdom*. Berlin: Springer Verlag, pp. 84–90.

Levy, A., Goren, M. G., Yosef, I., Auster, O., Manor, M., Amitai, G., Edgar, R., Qimron, U., and Sorek, R. 2015. CRISPR adaptation biases explain preference for acquisition of foreign DNA. *Nature* 520: 505–510.

Levy-Sakin, M., Pastor, S., Mostovoy, Y., Li, L., Leung, A. K., McCaffrey, J., Young, E., Lam, E. T., Hastie, A. R., Wong, K. H., Chung, C. Y., Ma, W., Sibert, J., Rajagopalan, R., Jin, N., Chow, E. Y. C., Chu, C., Poon, A., Lin, C., Naguib, A., Wang, W.-P., Cao, H., Chan, T.-F., Yip, K. Y., Xiao, M., and Kwok, P.-Y. 2019. Genome maps across 26 human populations reveal population-specific patterns of structural variation. *Nature Communications* 10: 1025.

Lewontin, R. C. 1997. Billions and billions of demons. *New York Review of Books* 44(1): 28–32.

Lewontin, R. C., Rose, S. P., and Kamin, L. J. 1984. *Not in Our Genes: Biology, Ideology, and Human Nature*. New York: Pantheon.

Liang, P., Xu, Y., Zhang, X., Ding, C., Huang, R., Zhang, Z., Lv, J., Xie, X., Chen, Y., Li, Y. Sun, Y., Bai, Y., Songyang, Z., Ma, W., Zhou, C., and Huang, J. 2015. CRISPR/Cas9-mediated gene editing in human tripronuclear zygotes. *Protein & Cell* 6: 363–372.

Löwy, I. 2017. Leaking containers: Success and failure in controlling the mosquito *Aedes aegypti* in Brazil. *American Journal of Public Health* 107: 517–524.

Ludmerer, K. M. 1972. *Genetics and the American Society: A Historical Appraisal*. Baltimore: Johns Hopkins University Press.

Lynch, M. 2007. *The Origins of Genome Architecture*. Sunderland, MA: Sinauer.

Maben, A. J. 2016. The CRISPR fantasy: Flaws in current metaphors of gene-modifying technology. *Inquiries Journal* 8(6). http://www.inquiriesjournal.com/articles/1422/the-crispr-fantasy-flaws-in-current-metaphors-of-gene-modifying-technology. Last accessed 17 August 2019.

Makarova, K. S., Aravind, L., Grishin, N. V., Rogozin, I. B., and Koonin, E. V. 2002. A DNA repair system specific for thermophilic archaea and bacteria predicted by genomic context analysis. *Nucleic Acids Research* 30: 482–496.

Makarova, K. S., Aravind, L., Wolf, Y. I., and Koonin, E. V. 2011b. Unification of Cas protein families and a simple scenario for the origin and evolution of CRISPR-Cas systems. *Biology Direct* 6: 38.

Makarova, K. S., Grishin, N. V., Shabalina, S. A., Wolf, Y. I., and Koonin, E. V. 2006. A putative RNA-interference-based immune system in prokaryotes: Computational analysis of the predicted enzymatic machinery, functional analogies with eukaryotic RNAi, and hypothetical mechanisms of action. *Biology Direct* 1: 7.

Makarova, K. S., Haft, D. H., Barrangou, R., Brouns, S. J., Charpentier, E., Horvath, P., Moineau, S., Mojica, F. J., Wolf, Y. I., Yakunin, A. F., Van Der Oost, J., and

Koonin, E. V. 2011a. Evolution and classification of the CRISPR-Cas systems. *Nature Reviews Microbiology* 9: 467.

Mali, P., Yang, L., Esvelt, K. M., Aach, J., Guell, M., DiCarlo, J. E., Norville, J. E., and Church, G. M. 2013. RNA-guided human genome engineering via Cas9. *Science* 339: 823–826.

Mand, C., Duncan, R. E., Gillam, L., Collins, V., and Delatycki, M. B. 2009. Genetic selection for deafness: The views of hearing children of deaf adults. *Journal of Medical Ethics* 35: 722–728.

Maroñas, O., Söchtig, J., Ruiz, Y., Phillips, C., Carracedo, Á., and Lareu, M. V. 2015. The genetics of skin, hair, and eye color variation and its relevance to forensic pigmentation predictive tests. *Forensic Science Review* 27: 13–40.

Marraffini, L. A., and Sontheimer, E. J. 2008. CRISPR interference limits horizontal gene transfer in staphylococci by targeting DNA. *Science* 322: 1843–1845.

Maxmen, A. 2015. The genesis engine, 22 July. http://www.wired.com/2015/07/crispr-dna-editing-2/. Last accessed 17 August 19.

McCabe, J., and Dunn, A. M. 1997. Adaptive significance of environmental sex determination in an amphipod. *Journal of Evolutionary Biology* 10: 515–527.

McClellan, J., and King, M. C. 2010. Genetic heterogeneity in human disease. *Cell* 141: 210–217.

McEvoy, B. P., and Visscher, P. M. 2009. Genetics of human height. *Economics and Human Biology* 7: 294–306.

Mojica, F. J., Díez-Villaseñor, C., García-Martínez, J., and Almendros, C. 2009. Short motif sequences determine the targets of the prokaryotic CRISPR defence system. *Microbiology* 15: 733–740.

Mojica, F. J. M., Díez-Villaseñor, C., García-Martínez, J., and Soria, E. 2005. Intervening sequences of regularly spaced prokaryotic repeats derive from foreign genetic elements. *Journal of Molecular Evolution* 60: 174–182.

Mojica, F. J., Díez-Villaseñor, C., Soria, E., and Juez, G., 2000. Biological significance of a family of regularly spaced repeats in the genomes of Archaea, Bacteria and mitochondria. *Molecular Microbiology* 36: 244–246.

Mojica, F. J. M., Ferrer, C., Juez, G., and Rodríguez-Valera, F. 1995. Long stretches of short tandem repeats are present in the largest replicons of the Archaea *Haloferax mediterranei* and *Haloferax volcanii* and could be involved in replicon partitioning. *Molecular Microbiology* 17: 85–93.

Mojica, F. J. M., Juez, G., and Rodríguez-Valera, F. 1993. Transcription at different salinities of *Haloferax mediterranei* sequences adjacent to partially modified *PstI* sites. *Molecular Microbiology* 9: 613–621.

Mojica, F. J. M., and Rodríguez-Valera, F. 2016. The discovery of CRISPR in archaea and bacteria. *FEBS Journal* 283: 3162–3169.

Monod, J., Changeux, J. P., and Jacob, F. 1963. Allosteric proteins and cellular control systems. *Journal of Molecular Biology* 6: 306–329.

Monod, J., and Jacob, F. 1961. Genetic regulatory mechanisms in the synthesis of proteins. *Journal of Molecular Biology* 3: 318–356.

Morange, M. 2015a. What history tells us XXXVII. CRISPR-Cas: The discovery of an immune system in prokaryotes. *Journal of Biosciences* 40: 221–223.

Morange, M. 2015b. What history tells us XXXIX. CRISPR-Cas: From a prokaryotic immune system to a universal genome editing tool. *Journal of Biosciences* 40: 829–832.

Moreno-Madriñán, M. J., and Turell, M. 2018. History of mosquito-borne diseases in the United States and implications for new pathogens. *Emerging Infectious Diseases* 24: 821–826.

Mugny, G., and Carugati, F. 1989. *Social Representations of Intelligence*. Cambridge, UK: Cambridge University Press.

Müller-Wille, S. 2014. Race and history: Comments from an epistemological point of view. *Science, Technology, & Human Values* 39: 597–606.

Murdock, G. P. 1949. *Social Structure*. Oxford, UK: Macmillan.

Nakata, A., Amemura, M., and Makino, K. 1989. Unusual nucleotide arrangement with repeated sequences in the Escherichia coli K-12 chromosome. *Journal of Bacteriology* 171: 3553–3556.

National Academies of Sciences, Engineering, and Medicine. 2017. *Human Genome Editing: Science, Ethics, and Governance*. Washington, DC: National Academies Press.

National Academy of Sciences. 2020. *Heritable Human Genome Editing*. Washington, DC: The National Academies Press. https://doi.org/10.17226/25665. Last accessed 17 November 2020.

Nelkin, D., and Tancredi, L. 1994. *Dangerous Diagnostics: The Social Power of Biological Information*. Chicago: University of Chicago Press.

Noble, C., Min, J., Olejarz, J., Buchthal, J., Chavez, A., Smidler, A. L., DeBenedictis, E. A., Church, G. M., Nowak, M. A., and Esvelt, K. M. 2019. Daisy-chain gene drives for the alteration of local populations. *Proceedings of the National Academy of Sciences (USA)* 116: 8275–8282.

Normile, D. 2019. Government report blasts creator of CRISPR twins. *Science* 363: 328.

Ohno, S. 1972. So much "junk" DNA in our genome. *Brookhaven Symposia in Biology* 23: 366–370.

Ouagrham-Gormley, S. B., and Fye-Marnien, S. R. 2019. Is CRISPR a security threat? In Singh, S. K., and Kuhn, J. H., Eds. *Defense Against Biological Attacks*. Cham: Springer, pp. 233–251.

Parens, E., and Asch, A., Eds. 2000. *Prenatal Testing and Disability Rights*. Washington, DC: Georgetown University Press.

Paul, D. B. 1998. *Politics of Heredity: Essays on Eugenics, Biomedicine, and the Nature-Nurture Debate*. Albany, NY: State University of New York Press.

Paul, D. B., and Brosco, J. P. 2013. *The PKU Paradox: A Short History of a Genetic Disease*. Baltimore, MD: Johns Hopkins University Press.

Pauling, L., Itano, H. A., Singer, S. J., and Wells, I. C. 1949. Sickle cell anemia, a molecular disease. *Science* 110: 543–548.

Pavan, W. J., and Sturm, R. A. 2019. The genetics of human skin and hair pigmentation. *Annual Review of Genomics and Human Genetics* 20: 41–72.

Pellicer, J., Fay, M. F., and Leitch, I. J. 2010. The largest eukaryotic genome of them all? *Botanical Journal of the Linnaean Society* 164: 1–15.

Penrose, L. 1946. Phenylketonuria: A problem in eugenics. *Lancet* 247: 949–953.

Perlman, R. L., and Govindaraju, D. R. 2016. Archibald E. Garrod: The father of precision medicine. *Genetics in Medicine* 18: 1088–1089.

Piaggio, A. J., Segelbacher, G., Seddon, P. J., Alphey, L., Bennett, E. L., Carlson, R. H., Friedman, R. M., Kanavy, D., Phelan, R., Redford, K. H., Rosales, M., Slobodian, L., and Wheeler, K. 2017. Is it time for synthetic biodiversity conservation? *Trends in Ecology and Evolution* 32: 97–107.

Plomin, R., and von Stumm, S. 2018. The new genetics of intelligence. *Nature Reviews Genetics* 19: 148–159.

Pourcel, C., Salvignol, G., and Vergnaud, G. 2005. CRISPR elements in *Yersinia pestis* acquire new repeats by preferential uptake of bacteriophage DNA, and provide additional tools for evolutionary studies. *Microbiology* 151: 653–663.

President's Commission for the Study of Ethical Problems in Medicine and Biomedical and Behavioral Research. 1982. *Splicing Life*. Springfield, VA: National Technical Information Service.

Quackenbush, J. 2011. *The Human Genome: The Book of Essential Knowledge*. Watertown, MA: Imagine.

Ramaswamy, S. 2017. Reversing pollinator decline is key to feeding the future. Blog. United States Department of Agriculture. https://www.usda.gov/media/blog/2016/06/24/reversing-pollinator-decline-key-feeding-future. Last accessed 12 December 2019.

Rana, P., and Craymer, L. 2018. Big tongues and extra vertebrae: The unintended consequences of animal gene editing. *Wall Street Journal*, 14 December. https://www.wsj.com/articles/deformities-alarm-scientists-racing-to-rewrite-animal-dna-11544808779. Last accessed 10 July 2019.

Reardon, S. 2019. Gene edits to 'CRISPR babies' might have shortened their life expectancy. *Nature* 570: 16.

Reilly, P. R. 1987. Involuntary sterilization in the United States: A surgical solution. *Quarterly Review of Biology* 62: 153–170.

Reilly, P. R. 2015. Eugenics and involuntary sterilization: 1907–2015. *Annual Review of Genomics and Human Genetics* 16: 351–368.

Rheinberger, H. J. 2000. Beyond nature and culture: modes of reasoning in the age of molecular biology and medicine. In *Living and Working with the New Mical Technologies: Intersections of Inquiry*. Cambridge, UK: Cambridge University Press, pp. 19–30.

Richardson, K. 2000. *The Making of Intelligence*. New York City: Columbia University Press.

Richter, J. 1991. Medicine and politics in Soviet-German relations in the 1920s: A contribution to Lenin's pathobiography. In Fierens, E., Tricot, J., Appleboom, T., and Thier, M., Eds. *Proceedings of the XXXIIth International Congress on the History of Medicine: Antwerp*, 3–7 September 1990. Bruxelles: Societas Belgica Historiae Medicinae, pp. 1063–1071.

Ridley, M. 2008. Foreword. In Witkowski, J. A., and Inglis, J. R., Eds. *Davenport's Dream: 21st Century Reflections on Heredity and Eugenics*. Cold Spring Harbor, NY: Cold Spring Harbor Press, pp. ix–xi.

Rieger, R., and Michaelis, A. 1954. *Genetisches und Cytogenetisches W¨orterbuch*. Berlin: Springer.

Ringman, J. M. 2007. The Huntington disease of Woody Guthrie: Another man done gone. *Cognitive and Behavioral Neurology* 20: 238–243.

Robertson, J. 1994. *Children of Choice: Freedom and the New Reproductive Technologies*. Princeton, NJ: Princeton University Press.

Rode, N. O., Estoup, A., Bourguet, D., Courtier-Orgogozo, V., and Débarre, F. 2019. Population management using gene drive: Molecular design, models of spread dynamics and assessment of ecological risks. *Conservation Genetics* 20: 671–690.

Rosenberg, N. A., Edge, M. D., Pritchard, J. K., and Feldman, M. W. 2019. Interpreting polygenic scores, polygenic adaptation, and human phenotypic differences. *Evolution, Medicine, and Public Health* 2019: 26–34.

Rossant, J. 2018. Gene editing in human development: Ethical concerns and practical applications. *Development* 145(16). doi:10.1242/dev.150888.

Rouet, P., Smih, F., and Jasin, M. 1994. Introduction of double-strand breaks into the genome of mouse cells by expression of a rare-cutting endonuclease. *Molecular and Cellular Biology* 14: 8096–8106.

Ruden, D. M., Bolnick, A., Awonuga, A., Abdulhasan, M., Perez, G., Puscheck, E. E., and Rappolee, D. A. 2018. Effects of gravity, microgravity or microgravity simulation on early mammalian development. *Stem Cells and Development* 27: 1230–1236.

Saleeby, C. W. 1909. *Parenthood and Race Culture: An Outline of Eugenics*. London: Cassell.

Santelli, R. 2012. *This Land is your Land: Woody Guthrie and the Journey of an American Folk Song*. Philadelphia: Running Press.

Sapranauskas, R., Gasiunas, G., Fremaux, C., Barrangou, R., Horvath, P., and Siksnys, V. 2011. The Streptococcus thermophilus CRISPR/Cas system provides immunity in *Escherichia coli*. *Nucleic Acids Research* 39: 9275–9282.

Sariola, S., and Gilbert, S. F. 2020. Toward a symbiotic perspective on public health: Recognizing the ambivalence of microbes in the anthropocene. *Microorganisms* 8: 746.

Sarkar, S. 1992. Para qué sirve el proyecto Genoma Humano. *La Jornada Semanal* 180: 29–39.

Sarkar, S. 1996. Biological information: A skeptical look at some central dogmas of molecular biology. In Sarkar, S., Ed. *Philosophy and History of Molecular Biology: New Perspectives*. Dordrecht: Kluwer, pp. 187–231.

Sarkar, S. 1998. *Genetics and Reductionism*. Cambridge, UK: Cambridge University Press.

Sarkar, S. 1999. From the *Reaktionsnorm* to the adaptive norm: The norm of reaction, 1909–1960. *Biology and Philosophy* 14: 235–252.

Sarkar, S. 2012. *Environmental Philosophy: From Theory to Practice*. Chichester: Wiley-Blackwell.

Sarkar, S. 2015. The genomic challenge to adaptationism. *British Journal for the Philosophy of Science* 66: 505–536.

Sarkar, S. 2018. Researchers hit roadblocks with gene drives. *BioScience* 68: 474–480.

Sarkar, S., and Gardner, L. 2016. Zika: The price of neglect. *Palgrave Communications* 2: 16060. doi:10.1057/palcomms.2016.60.

Sarkar, S., and Tauber, A. I. 1991. Fallacious claims for HGP. *Nature* 353: 691.

Savulescu, J. 2005. New breeds of humans: The moral obligation to enhance. *Ethics, Law and Moral Philosophy of Reproductive Biomedicine* 1: 36–39.

Savulescu, J., Pugh, J., Douglas, T., and Gyngell, C. 2015. The moral imperative to continue gene editing research on human embryos. *Protein & Cell* 6: 476–479.

Schermer, M., and Bolt, I. 2011. What's in a name? ADHD and the gray area between treatment and enhancement. In Savulescu, J., ter Meulen, S. R., and G. Kahane, Eds. *Enhancing Human Capacities*. Oxford: Wiley-Blackwell, pp. 179–193.

Schneider, V. A., Graves-Lindsay, T., Howe, K., Bouk, N., Chen, H. C., Kitts, P. A., Murphy, T. D., Pruitt, K. D., Thibaud-Nissen, F., Albracht, D., Fulton, R. S., Kremitzki,M., Magrini, V., Markovic, C., McGrath, S., Steinberg, K. M., Auger, K., Chow, C., Collins, J., Harden, G., Hubbard, T., Pelan, S., Simpson, J. T., Threadgold, G., Torrance, J., Wood, J. M., Clarke, L., Koren, K., Boitano, M., Peluso, P., Li, H., Chin, C.-S., Phillippy, A. M., Durbin, R., Wilson, R. K., Flicek, P., Eichler, E. E., and Church, D. M. 2017. Evaluation of GRCh38 and de novo haploid genome assemblies demonstrates the enduring quality of the reference assembly. *Genome Research* 27: 849–864.

Schork, N. J., Murray, S. S., Frazer, K. A., and Topol, E. J. 2009. Common vs. rare allele hypotheses for complex diseases. *Current Opinion in Genetics and Development* 19: 212–219.

Schwartz, M. 2018. Target, delete, repair: CRISPR is a revolutionary gene-editing tool, but it's not without risk. *Stanford Medicine (Winter)*. https://stanmed.stanford.edu/2018winter/CRISPR-for-gene-editing-is-revolutionary-but-it-comes-with-risks.html. Last accessed 23 November 2020.

Shaw, J. 2016. Editing an end to malaria. *Harvard Magazine*, May–June. http://www.harvardmagazine. com/2016/05/editing-an-end-to-malaria. Last accessed 29 November 2019.

Sherman, R. M., Forman, J., Antonescu, V., Puiu, D., Daya, M., Rafaels, N., Boorgula, M. P., Chavan, S., Vergara, C., Ortega, V. E., Levin, A. M., Eng, C., Yazdanbakhsh, M., Wilson, J. G., Marrugo, J., Lange, L. A., Williams, L. K., Watson, H., Ware, L. B., Olopade, C. O., Olopade, O., Oliveira, R. R., Ober, C., Nicolae, D. L., Meyers, D. A., Mayorga, A., Knight-Madden, J., Hartert, T., Hansel, N. N., Foreman, L. G., Ford, J. G., Faruque, M. U., Dunston, G. M., Caraballo, L., Burchard, E. G., Bleecker, E., R., Araujo, M. I., Herrera-Paz, E. F., Campbell, M., Foster, C., Taub, M. A., Beaty, T. H., Ruczinski, I., Mathias, R. A., Barnes, K. C., and Salzberg, S. L. 2019. Assembly of a pan-genome from deep sequencing of 910 humans of African descent. *Nature Genetics* 51: 30–35.

Sinsheimer, R. L. 1969. Prospect for designed genetic change. *American Scientist* 57: 134–142.

Smithies, O., Gregg, R. G., Boggs, S. S., Koralewski, M. A., and Kucherlapati, R. S. 1985. Insertion of DNA sequences into the human chromosomal β-globin locus by homologous recombination. *Nature* 317: 230–234.

Soyfer, V. N. 1984. *Lysenko and the Tragedy of Soviet Science*. New Brunswick, NJ: Rutgers University Press.

Sparrow, R. 2011. A not-so-new eugenics: Harris and Savulescu on human enhancement. *Hastings Center Report* 41: 32–42.

Sparrow, R. 2019. Yesterday's child: How gene editing for enhancement will produce obsolescence—And why it matters. *American Journal of Bioethics* 19: 6–15.

Spengler, O. 1993. *Lenin's Brain*. London: Penguin Books.

Stein, L. 2008. Congress passes bill barring genetic discrimination: Action culminates more than a dozen years of legislative haggling. https://www.scientificamerican.com/article.cfm?id=bill-bars- genetic-discrimination. Last accessed 13 January 2014.

Stevens, T., and Newman, S. 2019. *Biotech Juggernaut: Hope, Hype, and Hidden Agendas of Entrepreneurial Bioscience*. New York: Routledge.

Stolberg, S. G. 1999. The biotech death of Jesse Gelsinger. *New York Times Magazine*, 28 November, pp. 136–140, 149–150. https://www.nytimes.com/1999/11/28/magazine/the-biotech-death-of-jesse- gelsinger.html. Last accessed 25 June 2020.

Strasser, B. J. 1999. Sickle cell anemia, a molecular disease. *Science* 286: 1488–1490.

Sturm, R. A., and Larsson, M. 2009. Genetics of human iris colour and patterns. *Pigment Cell and Melanoma Research* 22: 544–562.

Swetlitz, I. 2016. College students try to hack a gene drive—And set a science fair abuzz. *STAT News*. https://www.statnews.com/2016/12/14/gene-drive-students-igem. Last accessed 12 November 2019.

Szostak, J. W., Orr-Weaver, T. L., Rothstein, R. J., and Stahl, F. W. 1983. The double-strand-break repair model for recombination. *Cell* 33: 25–35.

Tang, L., Zeng, Y., Du, H., Gong, M., Peng, J., Zhang, B., Lei, M., Zhao, F., Wang, W., Li, X., and Liu, J. 2017. CRISPR/Cas9-mediated gene editing in human zygotes using Cas9 protein. *Molecular Genetics and Genomics* 292: 525–533.

Tatum, E. L. 1966. Molecular biology, nucleic acids and the future of medicine. *Perspectives in Biology and Medicine* 10: 19–52.

Tauber, A. I., and Sarkar, S. 1992. The human genome project: Has blind reductionism gone too far? *Perspectives in Biology and Medicine* 35: 220–235.

Tauber, A. I., and Sarkar, S. 1993. The ideology of the human genome project. *Journal of the Royal Society of Medicine* 86: 537–540.

Texas Department of State Health Services. 2018. All Texas newborns are screened for these disorders. https://www.dshs.texas.gov/newborn/screened_disorders.shtm. Last accessed 1 November 18.

Timoféeff-Ressovsky, H. A., and Timoféeff-Ressovsky, N. W. 1926. Über das phänotypische Manifestieren des Genotyps. II. Über idio-somatische Variationsgruppen bei *Drosophila funebris*. *Wilhelm Roux' Archiv für Entwicklungsmechanik der Organismen* 108: 148–170.

United Nations Educational, Scientific and Cultural Organization. 1952. *The Race Concept: Results of an Inquiry*. Paris: UNESCO.

University College London. 2021. UCL makes formal public apology for its history and legacy of eugenics. https://www.ucl.ac.uk/news/2021/jan/ucl-makes-formal-public-apology-its-history-and-legacy-eugenics. Last accessed 20 January 2021.

Urnov, F. D. 2018. Genome editing BC (before CRISPR): Lasting lessons from the "old testament." *CRISPR Journal* 1: 34–46.

van der Weele, C. 1995. *Images of Development: Environmental Causes in Ontogeny*. Utrecht: Elinkwijk.

Vogel, K. M., and Ouagrham-Gormley, S. B. 2018. Anticipating emerging biotechnology threats: A case study of CRISPR. *Politics and the Life Sciences* 37: 203–219.

Vogt, O. 1926. Psychiatrisch wichtige Tatsachen der zoologisch-botanischen Systematik. *Zeitschrift für die gesamte Neurologie und Psychiatrie* 101: 805–832.

Wang, D., Tai, P. W. L., and Gao, G. 2019. Adeno-associated virus vector as a platform for gene therapy delivery. *Nature Reviews Drug Discovery* 18: 358–378. https://doi.org/10.1038/s41573-019-0012-9.

Wang, J. H., Wang, R., Lee, J. H., Iao, T. W., Hu, X., Wang, Y. M., Tu, L. L., Mou, Y., Zhu, W. L., He, A. Y., Zhu, S. Y., Cao, D., Yang, L., Tan, X. B., Zhang, Q., Liang, G. L., Tang, S. M., Zhou, Y. D. Feng, L. J., Zhan, L. J., Tian, N. N., Tang, M. J., Yang, Y. P., Riaz, M., Wijngaarden, P. V., Dusting, G. J., Liu, G. S., and He, Y. 2017. Public attitudes toward gene therapy in China. *Molecular Therapy Methods & Clinical Development*: 40–42.

Watson, J. D. 2008. Genes and politics. In Witkowski, J. A., and Inglis, J. R., Eds. *Davenport's Dream: 21st Century Reflections on Heredity and Eugenics*. Cold Spring Harbor, NY: Cold Spring Harbor Press, pp. 1–34.

Wheeler, D. A., Srinivasan, M., Egholm, M., Shen, Y., Chen, L., McGuire, A., He, W., Chen, Y.-J., Makhijani, V., Roth, G. T., Gomes, X., Tartaro, K., Niazi, F., Turcotte, C. L., Irzyk, G. P., Lupski, J. R., Chinault, C., Song, X.-Z., Liu, Y., Yuan, Y., Nazareth, L., Qin, X., Muzny, D. M., Margulies, M., Weinstock, G. M., Gibbs, R. A., and Rothberg, J. M. 2008. Complete genome of an individual by massively parallel DNA sequencing. *Nature* 452: 872–876.

White, D., and Rabago-Smith, M. 2011. Genotype-phenotype associations and human eye color. *Journal of Human Genetics* 56: 5–7.

Wirth, T., Parker, N., and Yl¨a-Herttuala, S. 2013. History of gene therapy. *Gene* 525: 162–169.

Witkowski, J. A., and Inglis, J. R., Eds. 2008. *Davenport's Dream: 21st Century Reflections on Heredity and Eugenics*. Cold Spring Harbor, NY: Cold Spring Harbor Press.

Wolff, J. A., and Lederberg, J. 1994. A history of gene transfer and therapy. In Wolff, J. A., Ed. *Gene Therapeutics: Methods and Applications of Direct Gene Transfer*. Boston: Birkhäuser, pp. 3–25.

Wolstensholme, G., Ed. 1962. *Man and His Future: A CIBA Foundation Volume*. London: Churchill.

Zentner, G. E., and Wade, M. J. 2017. The promise and peril of CRISPR gene drives. *BioEssays* 39: 1700109.

Index

About the Author

Sahotra Sarkar is an Indian-American academic, author, and political critic. Originally from Darjeeling, he was educated at Columbia University and the University of Chicago where he obtained a PhD in philosophy in 1989. He was a leader of the student anti-apartheid movement in the 1980s and also worked with the African National Congress in southern Africa. As a critic of creationism he was active against efforts to infuse religion into science classes in the 2000s. During that period he published *Doubting Darwin? Creationist Designs on Evolution* (Blackwell, 2007) which was a scientific and philosophical critique of Intelligent Design creationism. For the last thirty years he has also been active in efforts to conserve biodiversity through local initiatives in various parts of the world including the Eastern Himalaya, the southern United States, and Mesoamerica. He co-authored *Systematic Conservation Planning* (Cambridge, 2007) with Chris Margules. He is the author of five other books and more than 250 scientific and philosophical articles. He has previously taught at Boston University and McGill University. He has been a Fellow of the Wissenschaftskolleg zu Berlin, the Dibner Institute for the History of Science (MIT), and the Edelstein Centre for the Philosophy of Science (Hebrew University). He has also held long-term visiting appointments at the Indian Institute of Science (Bangalore), the Museum of Contemporary Zoology (Harvard), the Max Planck Institute for the History of Science (Berlin), and the University of New South Wales (Sydney). He is currently Professor of Philosophy and of Integrative Biology at the University of Texas and lives in Austin with his wife and two daughters.

CPSIA information can be obtained
at www.ICGtesting.com
Printed in the USA
BVHW040258270721
612774BV00001B/2